# The Human Mind Revealed

This book provides readers with a new working model of the human mind based on contemporary trends in cognitive science. Taking a distinct biological approach, the book presents mind and brain as two sides of the same coin while at the same time emphasising the need for separate accounts. The reader is introduced step by step to the mind account, showing how it operates.

The text explores the mind's basic architecture, examining how its family of 11 systems interact flexibly and collaboratively to process information, mostly subconsciously. Key topics include the five systems that build specialised representations from sensory input, consciousness, memory, emotions, comparisons between human and animal cognition, language processing and concepts of self, free will and spirituality. The book illuminates how general principles of activation, association and competition act in combination with the unique principles that govern each of the mind's specialised systems. One chapter explains two of these systems in more detail to illustrate how expertise from any relevant area of research can be used to elaborate the basic model.

Written in an accessible style, *The Human Mind Revealed* offers valuable insights for students and scholars across cognitive science disciplines as well as general readers curious about how the mind works.

**Michael Sharwood Smith** is Emeritus Professor of Languages at Heriot-Watt University in Scotland, UK. His major research interest is cognitive representation, processing and development with special reference to human communication and the language sciences.

# The Human Mind Revealed
A Biological Perspective

Michael Sharwood Smith

Routledge
Taylor & Francis Group

LONDON AND NEW YORK

Designed cover image: Getty Images

First published 2026
by Routledge
4 Park Square, Milton Park, Abingdon, Oxon OX14 4RN

and by Routledge
605 Third Avenue, New York, NY 10158

*Routledge is an imprint of the Taylor & Francis Group, an informa business*

For Product Safety Concerns and Information please contact our EU representative GPSR@taylorandfrancis.com. Taylor & Francis Verlag GmbH, Kaufingerstraße 24, 80331 München, Germany.

*British Library Cataloguing-in-Publication Data*
A catalogue record for this book is available from the British Library

ISBN: 978-1-032-97129-2 (hbk)
ISBN: 978-1-032-99886-2 (pbk)
ISBN: 978-1-003-60653-6 (ebk)

DOI: 10.4324/9781003606536

Typeset in Sabon
by SPi Technologies India Pvt Ltd (Straive)

# Contents

*Preface*                                          *vii*

1   Minding the Brain                               1

2   Brain and Mind Maps Compared                    9

3   The Knowledge Network                          18

4   Reflections of Reality                         28

5   Building our Inner World                       37

6   Activating Knowledge                           50

7   Remembering and Forgetting                     63

8   Making Meanings                                75

9   Value and Emotion                              85

10  Space and Movement                             99

11  Comparing Species                             107

12  Conceptualising Consciousness                 115

13  Different Modes of Awareness                     125

14  Communication between Humans                 134

15  Human Language Decoded                         142

16  Concluding Reflections                          157

      *Bibliography*                                   *168*
      *Short Glossary*                                 *173*
      *Index*                                          *177*

# Preface

This book is about the human mind. The 'revealing' announced in the title of this book has something to do with the brain and a lot to do with a ghost. In the media, a great deal of attention these days is devoted to the human brain. In this general fascination with reported advances in brain science, a fascination which I share, there is still one distinction that has become blurred. Questions include: 'How is the mind related to the brain?' and 'What is the mind anyway?' Without any clarification, answers to such questions tend to switch first to the more tangible of these two concepts, away from the object of psychological investigation to what neuroscientists investigate. This expresses a need for something that can be seen, weighed, measured directly and located. Reflecting this preference for the more tangible concept, books purportedly on the mind often contain discussions more about the neurons involved. The main aim of the current book is to reset this state of affairs. Accordingly, it will include a detailed discussion of how the mind is organised and how the mind operates in our daily lives. And importantly, it will also spell out what it is and what its relationship with the brain is.

In the chapters that follow, the human mind will be discussed as a biological phenomenon, that is to say, no less a part of biology than the central nervous system. The book, you can say, is an introduction to our biological 'software'. Despite passing references to neurons, the spotlight will always be on the biological principles and systems that drive our actions, our thoughts and feelings.

The mind/brain distinction can, of course, be explained (or dismissed) in different ways. Some people think of the mind as an entirely separate and independent entity, perhaps not open to scientific explanation at all. Psychologists might not always agree, but the philosopher, Gilbert Ryle, in a critical mood, referred to the mind as the 'ghost in the machine',[1] dismissing 'dualist' ideas that treat the mind as an entity or 'presence' outside the body but still somehow mysteriously controlling it. This book aims to banish Ryle's ethereal ghost, bring the mind into the spotlight and introduce it as something every bit as real as the brain. The brain and mind in this perspective turn out to be two sides of the same coin, each meriting a separate type of description, but both are part of a comprehensive explanation of who we are. As such, they also represent two ways in which the story of human evolution can be told.

Acknowledgements are due to those who have inspired me to go 'where angels fear to tread' and dare to write a book such as this, not least including a host of researchers and thinkers who are too many to mention by name. This includes friends, colleagues and fellow researchers and also great pioneers who have contributed so much to our knowledge of different facets of the mind. I naturally need to mention John Truscott, fellow creator of the framework, without whom this account would not have been possible. His ideas pervade every page of this book, even those remarkably few aspects where we still maintain differing opinions, and his comments on the chapters of this book have been greatly valued. Special thanks are due also to Martha Obbink, who has played the much-appreciated role of a tireless general reader for me all along. I should also mention Ema Olaru, who, as a young, aspiring student of neuroscience, read earlier versions. Thanks too to occasional commentators like my long-term friend and colleagues, James Pankhurst and the late Paul van Buren, who queried my ideas and have often set me in the right direction or pointed to improvements in my explanations. I should also mention many relevant discussions with Ellen Bialystok and the many other researchers in various fields of cognitive science to whom I also owe a debt of gratitude, but are not identified by name. I hope, nonetheless, that they have all been suitably acknowledged in related academic publications bearing my name, either as author or coauthor. Finally, I acknowledge with gratitude the helpful comments, criticisms and suggestions of my anonymous reviewers.

## Note

1 Ryle, G. (1949).

# 1 Minding the Brain

## Introduction

Chapter 1 introduces a particular perspective on the nature of the human mind. It tackles questions like: 'what is your mind and where can I find it?' In the course of talking about the mind in general, a few other questions are also posed, such as 'Why do we need a separate account of the mind in the first place?' and 'How is the mind related to the brain?' Then, there is a related question about the nature of what we call 'knowledge'.

In this book, the mind will be treated as a *psychological* phenomenon having to do with thoughts, emotions and observable behaviour. Furthermore, since the mind is also the possession of living beings, in our case, human ones, it should also be treated as a *biological* phenomenon. This will be an important assumption throughout the book.

Knowledge will be treated, first and foremost, as the possession of *single individual minds*. This is irrespective of the ideas, facts and beliefs that get passed on second-hand and in some recorded form to other individuals and only then become a shared possession of other individuals. Finally, the human mind, as something that characterises our species, can be described and explained entirely in its own terms. This is despite the fact that its various states and activities are also reflected, albeit in different ways, in the physical brain. Put another way, the mind does not have neurons, and the brain does not have ideas. Nevertheless, the two together, both sides of one coin, form an inseparable unit.

It is also important to stress that knowledge is not only restricted to what we consciously know, think about and discuss with others.

DOI: 10.4324/9781003606536-1

A great deal of it lies beyond the reach of consciousness. Even so, this hidden knowledge can still strongly influence our thinking, feeling and behaviour. Furthermore, knowledge is generally discussed here as consisting of different types of cognitive *representation*. In most cases, it is the mind's way of interpreting different aspects of experience and storing its knowledge accordingly. One legitimate way of looking at what the mind's knowledge is all about and, in particular, the various ways in which it gets stored and used, is to view it as the brain's biological 'software'. Invisible but still very real. A very different computer software that has been deliberately designed but which, like the brain, is always evolving through time, shaped by natural, evolutionary processes, gradually adapting to the conditions under which we live. In this sense, its physical manifestation, the brain, is the 'mind's biological hardware'.

## Looking for the Mind

### *The Location Problem*

We all know where our brain is. It's a lump of living matter which sits safely inside our skull and weighs about 1.3 kilograms. That's roughly three pounds of wobbly soft matter – perhaps not a thing of beauty to the non-expert but truly awe-inspiring given its enormous complexity and what its odd 86 billion neurons do for us.[1] Thanks to modern technological progress, we are getting better and better at tracking and displaying the brain's activity. At the same time, as our most advanced brain scanners are now able to, ever more questions are raised about how its intricate patterns exactly relate to the combination of highly complex thoughts, emotions and observable behaviours that make up the mysterious 'you'. It is another Amazon jungle, another deep ocean world that science is still a very long way from fully understanding.

One problem with the mind is, since we cannot examine it directly, establishing how to get to grips with what it actually is. Apart from its obvious complexity, where do we locate it – in the head, for example? Or is that perhaps the wrong question? Perhaps the mind only exists, in the head or elsewhere, in a strictly metaphorical sense, although we have no problem talking about it in everyday language as though it is real. Despite the fact that, for many, the 'mind' remains an elusive concept, it is indeed very much part and parcel of our everyday thinking. It frequently occurs in our conversations: 'Are you out of your mind?', 'Have you lost your mind?', 'My mind's not working well this morning' or 'It's all in the mind': these are all everyday expressions in English. Note that only some of these mentions of the mind could be replaced by 'the brain'. We don't say 'You are out of your brain' or 'I'm going to make up my brain', but we might also say 'My brain's not working well this morning' since that metaphor seems to convey the idea of a machine which is not functioning properly. Not surprisingly, the brain and the mind sometimes seem to be used as two names for the same thing. I am certainly not the first or the last person to acknowledge this confusion of the two concepts, mind and brain, and think that the mind, especially in recent times, has not been given the attention it deserves. This book will now try to provide some concrete answers.

Whatever its relationship with the brain in our heads, for most of us the mind does appear to be the ultimate source of everything that makes us tick: how we feel, what we think and why we behave the way we do. Not surprisingly, then, it is a topic about which novelists and poets and, in a more strictly scientific manner, psychologists and numerous philosophers as well have been writing informatively for centuries. Many of the notions discussed in this book will certainly resonate with ideas and theories in philosophy as well as with deeply aesthetic and spiritual experiences that some have reported. If we still feel the need to ask the question *'where* is my mind, actually?', the doctor will probably not be of much help. It may seem to be a very abstract concept, but we do somehow feel it is real, so our attention naturally first turns to a location a few centimetres above our shoulders. It still feels right somehow to locate the mind where the brain happens to be, albeit in some vague, undefined way. By doing so, we are, by implication, treating the brain and the mind as related to one another. How, then, are we going to make the mind into something more concrete, something much easier to grasp at least at a basic level?

There are many interesting books that are presented as being about the brain but are actually very much about the mind. For instance, there is a book about emotions[2] by Lisa Feldman Barrett, a psychologist and neuroscientist. It has an interesting subtitle, which is relevant to the current enterprise. In it, she refers to 'the secret life of the brain'. That, for me, is a tempting alternative description of the mind, although the plan is now to make the mind a great deal less vague and a good deal less secret.

### The Brain is Missing Something

The tendency to go straight to the brain for getting all the answers to questions about the mind has been a feature of various questionable uses of the 'neuro-' term in the public domain with the term achieving an almost mystical status. In the approach to the mind adopted here, there can really be *biological* answers to questions about the mind without even breathing a word about neurons. That, of course, is not how things should remain. One day, as research into human behaviour, thoughts and feelings progresses, both brain and mind dimensions will become much more comprehensively integrated in scientific discussion. This has long been the ambition of many scientists, as Eric Kandel, a neuroscientist and a Nobel prize winner, explains in *The Age of Insight*,[3] his superb and highly readable account of how the scientific approach to mind and brain was born at the turn of the 20th century in his native city of Vienna. Freud, who had begun his career as a neuroscientist, produced in 1900 a detailed psychological theory of the mind (*The Interpretation of Dreams*[4]), separate from any manifestations in the brain, while William James, a trained physician, had produced his own approach to the mind a year earlier in his equally groundbreaking *The Principles of Psychology*.[5]

## The Need for a Guiding Framework

There are a number of things you can ask when trying to get a sense of the difference between brain and mind. One is about damage. The brain, of course, can sustain physical damage as a result of an accident. This does not *necessarily* mean damage to the mind, but it can. However, since the relationship between how your physical brain works and the way your mind works is complex and indirect, your work is going to be impossible if you confine your investigations to the brain alone. This is exactly why we need *separate* accounts for how the mind as a whole functions. To do this, at the very least, a framework is needed for describing the mind and how it works, a basic template which can then be elaborated by experts to different areas of mind research.

Since there exists no suitable, comprehensive theory of the human mind, a theoretical framework (already mentioned in the Preface will be used for the present account). What using a framework does is, instead of presenting you with a list of different topics about the mind with both controversial and established findings, it distils mutually compatible trends in contemporary science into a single, consistent, basic story about the mind's organisation and operating principles. In this way, a starting point or 'base camp' is provided from which more targeted explorations can be mounted. In this way, the theoretical framework can lead on to what would look like a much more elaborated *theory* of the mind. In the present context, it will serve as an initial account, but one that already reveals essentially what the mind is and how it works from a biological as well as a psychological perspective.

## The Two Rings

Before we get on to explaining how the mind works in any more detail, I will be using, as a guide, a map of something familiar, namely a simple city metro system. This will work as a useful metaphor and point of reference as we progress through the chapters.

Figure 1.1, in its simplest form, shows the map that we will be using. The details can wait until later.

The map in Figure 1.1 shows two rings, an *outer* one with five mental ('cognitive') systems, each dedicated to making their own interpretation of particular physical signals that originate in the outside environment and impact on our five senses. In addition, there is an *inner* ring with six further systems crucially needed for building a much richer, internal interpretation of reality, that is to say, as far as we are able to understand and respond to it.

These two rings, reshaped as rectangles with rounded corners, portray the mind as a kind of map display that passengers will see in any London Underground Station, for example. This mind is thus presented as a fixed network of 'expert systems': these are the mind's metro 'stations'. For example, one of them located on the inner ring, labelled 'A', runs your 'affective system'. The affective system handles the way your mind attaches positive and negative values as well as your basic emotions like fear and happiness. Another system on the outer ring labelled 'AU' (for 'auditory') tries to make sense and create representations of the sound wave patterns that strike your

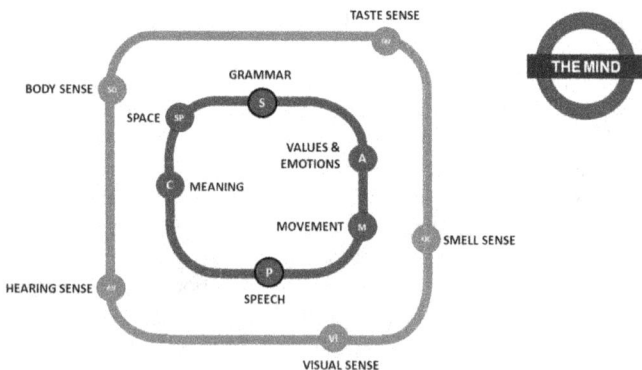

*Figure 1.1* Two rings to rule your mind.

ears. Such maps displayed in stations are not there to help the railway engineer or traffic controller but are designed for the benefit of the average user, the passenger who wants to navigate their way to some chosen destination. In the chapters that follow, this is more or less what I shall be trying to do with the metro map of the mind. There are details to be added, but for all the inevitable differences that will emerge between the human mind and a metro system, the metro map of the mind will serve as a handy reference point and maintain an idea of the mind as a set of interconnected systems that together form an integrated whole.

Metaphors, of course, can never be taken too far. As just mentioned, there are indeed a number of obvious ways in which a mind *differs* from a railway network. At the same time, there is perhaps one more comparison that we could still sneak in: and that is how the mind operates at different levels. For example, some of the mind's activities take place 'above ground'. These are the activities where we are engaged in thought, aware of our feelings and noticing what is around us. The rest of the mind's activities – like the parts of their metro system that Londoners refer to as 'the Tube' – operate underground, that is, *sub*consciously. In the case of the mind, this covers by far the largest part of its activities. Put another way, none of us has any real idea about what most of our minds are actually doing. An unsettling thought, perhaps, and a problem which will now need to be worked out more fully.

## Minds and Machines

### *Are We Just Machines, After All?*

There always remains that nagging cause of anxiety these days, at least for those who worry about whether minds and brains should be put on an equal footing with very advanced computers, for example, those controlling robots with artificial intelligence (AI). Creating a map of the mind like the one shown above might well suggest that it has been reduced to some sort of blueprint for a silicon chip or motherboard, one that might run a laptop or a washing machine. Just in case the map displayed in Figure 1.1 triggers the same suspicion, I should straightaway make it clear that the answer is no. The mind is much more than that as I hope to demonstrate in the following chapters.

## Minds as Supercomputers

Minds and brains may well suggest 'computers' in the sense that they are able to perform computations in order to solve problems. Minds can rapidly perform calculations and come up instantly with results. However, we are complex *biological* beings. Unlike such artefacts that humans create, minds and brains have *evolved* over millions of years. They have developed their own particular systems by interacting with and adapting to their environment. This means not only dealing with disease, deserts, tsunamis, climate changes and hostile wildlife but it also means interacting usefully, emotionally, productively and creatively with fellow members of the same species, including very hostile ones. The human mind's amazing biological software that enables all this has been fashioned over many hundreds of thousands of years by the processes of evolution.

You could describe the development of minds (and brains) over evolutionary time as a natural, slow design-and-adapt process. This rolls on and on through the millennia: very, very gradually, humans and all the other life forms change in a way that should optimise their survival in the particular environment in which they spend their lives. What is more, as each single individual human travels through their own particular life equipped with the same basic DNA as any other human being, their minds also grow, shaped by their own experiences. This makes each person and each person's *mind* increasingly different from anyone else's. So, even though all of us, as possessors of human minds, basically work in the same way, every mind turns out to be unique.

## Notes

1 Not forgetting the support and maintenance system consisting of about the same number of astrocytes and other glial cells.
2 Barrett, L. F. (2017).
3 Eric Kandel (2012).
4 Freud (1997). (A translation into English of the original *Die Traumdeutung*).
5 James (1890).

# 2 Brain and Mind Maps Compared

## Introduction

In this chapter, human vision is used as an example of how *any* of the mind's 11 systems can also be described, both in the mind's own terms but also in the very different terms of the brain. The five systems on the mind's outer ring use physical input coming from each of the five sense organs to build their own specialised representations or 'knowledge' of the world outside. Each of these five systems has a *processor* that manages only one type of signal, in this case, visual signals. Each also has its own *store* in which its own type of accumulated knowledge is located. This basic two-component architecture (store plus processor) is repeated in all 11 systems. The way we perceive what is outside us is then by *interpreting* it in five different ways. The resulting representations are created and stored in one or other of the five outer ring systems.

## A Scientific Approach

In referring to the mind as a biological phenomenon, it should be clear that the mind is not being treated as something mystical or spiritual like a 'soul'. If it were, very different kinds of explanation would be required, and these would involve beliefs and assumptions not yet supported by modern science. Without adopting such views as alternatives or even as supplementary enhancements of the present biological perspective, we are still left with something that is already wondrous, intriguing and meaningful. Both the mind and brain are part of the same amazing and exciting biological story of who we are.

DOI: 10.4324/9781003606536-2

Imagine a rapidly spinning coin on a flat surface: we see it as a single object, a shiny globe whirring around. If we stop it whirring, we discover a single coin with two sides to it. In my metaphor, one side is the brain side, and the other is the mind side. Each side is different, but both are sides of the same coin. Mind and brain are unified, part of a whole, but viewed and described from two different perspectives[1]. Perhaps the only qualification to be made here might be with reference to certain support systems and processes in the brain, in particular, the autonomic nervous system. This takes care of involuntary functions like breathing, heart rate and the sleep/wake cycle. These can indeed be affected by and themselves influence mental states, but here they will be mostly left aside as being non-essential for present purposes.

### The Example of Vision

Here, to illustrate the differences between mind and brain accounts, are two concrete examples of two different but related maps. The human *visual* system will be used. There is a neuroscientific (hardware) way of explaining it and also a parallel psychological (software) way. Light wave patterns are picked up in the eyes and converted into electrochemical signals. This is the point at which both accounts of visual processing diverge and become either the *neurological* (brain) explanation or the *psychological* (mind) one. Let us take the brain option first.

### Vision in the Brain

The first (already considerably simplified) map in Figure 2.1 deals with just part of the biological 'machinery' involved in visual activity that takes place in the brain. It shows the early stages of the whole process leading, in the mind, to the creation of our inner visual landscape.

Then, in simple terms, visual processing starts when patterns of light hit the back of each eye on surfaces called retinas. These patterns leave an imprint, which marks the first step in the processing chain, which then continues as a series of stages. The optic nerve carries the information forward, with pathways from each eye travelling along separate routes to end up in that part of the

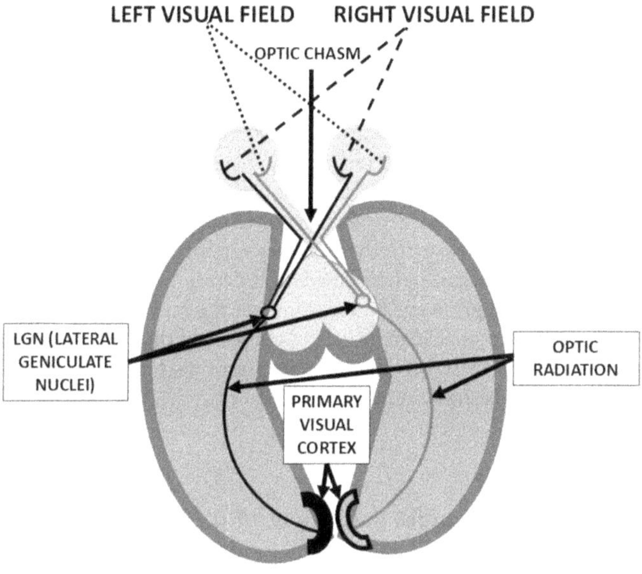

*Figure 2.1* Mapping the brain's visual system.

brain which contains the primary visual cortex. 'Optic radiation' on each side from the lateral geniculate nucleus (one on the left and one on the right) marks the final stage shown in the figure. The visual cortex (V1) is where your brain begins to interpret what your eyes have seen. Visual processing continues involving other areas of the visual cortex (V2–5) that are not shown. Indeed, much of the brain that is relevant to the current account of the mind concerns various parts of the cortex, the outer layer covers the largest part of the brain.

How our minds make sense of signals from the world around us – received through our five senses – depends on these initial steps being successful. In this example of visual perception, the light patterns detected by the retina must be processed in a way that the visual system can interpret them. Once this happens, the mind can begin the interpretation stage and make visual sense of the outside world. The entire flow of information normally happens in a split second.

These various details included on the brain map (in Figure 2.1) are all parts of the visual processing and storage in your skull. It is not necessary for current purposes to understand all the details of what happens in the brain. This map just gives a brief and very partial impression of the hardware nuts and bolts, the tunnels, wiring and cabling needed for processing the visual experience, together with some indications of the wave of activity that flows out from each eye during the actual processing of visual information.

Now, let's cross over to the software version. This is the way in which vision is organised in the *mind*. It determines how the mind responds after light wave patterns impact the retina. Setting aside what happens in *neural* matter, explanations of the resulting *mental* activity can be streamlined and its essentials set out in its simplest form, as is shown in Figure 2.2.

### Vision in the Mind

The visual system in the mind is one of the five systems on the outer ring of the mind's 'metro map' (see Figure 1.1). Each system deals with signals coming to us from the external world outside us. Figure 2.2 now provides the basics of the visual *software*. As can immediately be seen, the basic organisation of visual processing in

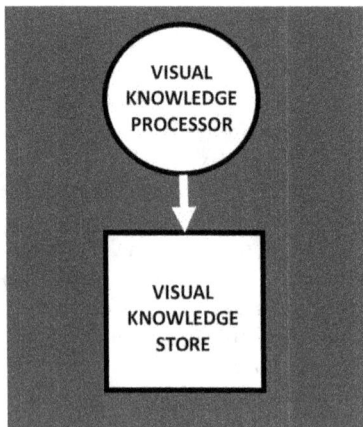

*Figure 2.2* The mind's visual system.

the mind can be reduced neatly to two basic components. This two-component system, in fact, reflects the basic make-up of *any* of the 11 expert systems, each specialised in a particular kind of knowledge. In the case of vision, the visual software has just two basic components for creating and managing the mind's visual knowledge and are labelled:

1 The Visual Processor
2 The Visual Store

The processors manage all the activity which they are uniquely specialised to handle. In this case, it happens to be the visual processor handling visual activity. The contents of the stores are what each specialised processor operates on and where the outcomes of processing activity are kept. The visual store contains what you might call processed 'knowledge' created on the basis of activity originating in the outside world impacting on the relevant sense organ. In this case, the environmental activity in question consists of light waves impacting the back of our eyes.

The crucial feature that differentiates each system from all its partners is the bunch of principles that its processor uses to encode, store and manage its particular type of knowledge. In other words, each system has, over and above *general* principles that apply to all *mental* processing, its own specialised operating principles. The *visual* processor in this case follows *visual* principles. These principles have evolved over time to help humans interpret their visual experience and build representations of this experience to store. Visual principles and the visual knowledge structures created when applying them are quite different from the principles that are applied in other systems. Each of these perceptual systems creates its own response to the raw signals that come from outside from sense organs other than the eyes. As is the case with all these five systems, the visual processor will, indeed *must*, automatically have a go and do its best to make sense of all incoming physical signals, whether or not the patterns they form can be effectively processed. All expert systems will automatically do their best in any given situation and work with the information they currently have available to them. Moreover, whatever the actual result, they do their creative work subconsciously without us being aware of it.

In a curious and indirect manner, this basic display of the mind's organisation as a network of specialised systems is able to do what Franz Joseph Gall could not do. Gall, born towards the end of the 18th century, is known as the founder of the science of 'phrenology', now considered to be a pseudoscience. He was convinced that different mental faculties could be traced to specific locations in the brain, and the size of these individual parts of the brain could be detected in the way the skull was shaped. His definitions of the various 'faculties' of the mind were flawed; their presumed locations were completely arbitrary, as were the connections between faculty size in the brain and skull contours, the visible bumps on the head. In rather the same way, there is no direct relationship between the metro map of the mind and the anatomical details of the brain. It would have been a serious error if that had been the claim.

By making a clear distinction between how the systems of the *mind* are neatly displayed as single locations (but not as physical bumps) and how equivalent specialised systems are physically manifested in *neural* matter, at least some element of Gall's basic idea can be retained. There is, in fact, also a growing trend in modern neuroscience to talk about distinct functionally specialised systems in the brain as well, even though such brain systems are a lot more complex than Gall imagined and not distributed so neatly in single locations.

To sum up the software account thus far, there are five specialised systems on the outer ring of the mind, each of them receiving signals (input) from outside. You can assume that there are also five different processors, each specialised in dealing with one kind of input, five distinct sets of operating principles for doing just that and finally five separate stores for storing five types of knowledge. The visual system was just one example of how this shared architecture works.

As far as how representations – our own internal 'knowledge' of the world – are built up from experience is concerned, what ends up in each of the different stores will still differ from individual to individual. This happens even though we all have the same set of systems, the same stores and the same operating principles for dealing with reality. There will be a sufficient amount of knowledge in common for humans to be able to share their experiences, but ultimately, no one individual is identical to any other, even 'identical twins'. This theme of individual differences will surface again in the chapters that follow.

## Building an Inner Version of Our Outer World

Because we have this sense of instant understanding of what is outside us, it is quite difficult to accept that what for us is 'the outside' is in fact an internal creation of our minds. On viewing a new object in the immediate environment, for example, an unfamiliar animal, the visual processor, with the help of its own special processing principles, creates what amounts to a 'representation' of that image in its visual store. It takes the form of a structure of varying complexity, which is a combination of other representations, including colours and shapes. Once created, this visual structure is immediately stored, like a memory on a flash drive. However, it is best thought of not as a memory but as a little chunk of visual knowledge, a representation. In any case, for as long as it remains in the visual store, and still easily accessible (more on that notion later), it can be called upon ('activated') as and when necessary. When that happens, we can also think of it as a visual memory. Note that this activation process means simultaneously activating all those individual visual component features that together combine to form that complex representation. This means that the representation of the object out there in the world outside us, like a pin, a light bulb, a face or a building, has been constructed by our mind's visual system by putting together its associated features and integrating them into a composite whole. All of these separate structures are assembled according to the principles of its processor. Our internally created visual world consists of all the visual structures (representations) that have been accumulated so far in our visual store, all ready to be activated in one way or another on any particular occasion.

Staying with our example of the visual system, once we or rather our mind's visual processor has constructed a representation of an unfamiliar object, it will be activated again whenever the object in question (or anything resembling it) has been encountered. At this point, the light waves emanate from what we might see as a single object and make contact with the retinas in our eyes. We then 'see' the object not as patterns on our retinas and certainly not as light waves, but as an *interpretation* of that information by the processor in our visual system. Our ongoing visual experience will be an interaction between our internal and external visual worlds. It will be based on a blend of already stored visual knowledge representations in our visual store, now activated, plus anything being currently processed in the visual scene outside us and in the process of being created internally. Lighting and the angle at which we are viewing the object, for instance, will certainly affect the way we perceive it at that particular moment and might even make it difficult to recognise immediately, although we can retain a stable notion of what the object should look like under different conditions.

As with anything else, we can talk about the brain's visual system or we can set aside the myriad physical details of our hardware and concentrate on what our biological software is doing. As we are now doing. Moreover, by looking at the basic way in which the mind works and excluding the physical detail, we can more easily see the visual system as an integrated part of a whole family of mind systems, all working at the same time and in the same basic manner.

As already indicated, one person's knowledge is unique because no individual's experience is the same as another's. Even similar experiences may be recorded and stored in different ways. Two people can see the same thing, say a dog, in front of them using their eyes but what gets processed and stored in their heads will not end up as perfect 100% copies of one another. In other words, they will be similar but not identical. Was that dog's tail just black, yellow and black or brown and yellow? Three individuals might argue about what they saw. What another animal makes of the same object in front of them will be naturally even *more* different, just because its brain and its mind work in markedly different ways.

Each system works in essentially the same way. What has been said here about the current example, the visual system, also applies to the auditory, olfactory (smell), gustatory (taste) and somatosensory (body sense) systems as well. Together, these five systems on the outer ring contribute to our combined sense of the physical world in which we live.

### The Mind as a Recycler

Whatever system we are talking about, the visual system or any other system on either ring, building a complex representation will always mean, in practice, combining much simpler representations but in different ways. It is like creatively constructing many different shapes with combinations of its own unique type of Lego brick. Perhaps even more important is the fact that, in making all these combinations, the mind reveals itself as a Great Recycler. As we shall see, 'new' representations are actually new *combinations* of representations that are already in the store. Elements of familiar dogs' 'tail' will be incorporated into the representation of newly encountered animals' tails. In the next chapter, we will go into this whole knowledge creation process in more detail.

### Specialisation, Connectivity and False Divisions

Finally, the dominant characteristic of both the brain and mind is specialisation. They both manage the tasks we have to perform using multiple systems, each of which has a special function. Specialisation, then, is one shared characteristic of brain and mind. However, when relating the two, you come across statements like 'language is on the left side, emotions on the right'. Language is a very broad and diverse phenomenon covering both sides of the brain. Emotion likewise involves locations and connections on both sides. This makes such simple divisions misleading. For example, you can distinguish *conscious* and *subconscious* mental activities but again, this does not mean that the mind itself is divided into two separate parts. Also, having a set of specialised systems does not mean they are completely isolated from one another. As the following chapters will make clear, if specialisation is one major characteristic of brain and mind organisation, then another one, one that provides a balanced overall picture of both systems, would be *rich interconnectivity*.

## Note

1 This 'dual-aspect' way of thinking about the mind/body relationship has a long history in philosophy and could be called dual aspect-monism as espoused by, for example, thinkers like Eric Kandel, Wolfgang Pauli, Carl Jung and to some extent, the phenomenologist Merleau-Ponty.

# 3   The Knowledge Network

## Introduction

The map of the mind introduced earlier showed the mind as a fixed network or 'family' of specialised systems arranged around an inner and an outer ring. However, this metro analogy has its limits: railway lines are for moving passengers between stations. The specialised knowledge in each system never moves from its store. It is never merged with another type of knowledge. Instead, a network of pathways links different stores through which web associations are formed. These associations associate knowledge of one type with other types. If one representation is activated, all of its associates are co-activated at the same time. These webs of association are called 'schemas'. These are the mind's way of recruiting its specialised systems flexibly to collaborate and solve a huge variety of problems.

Only a small percentage of the mind's activities involves conscious processing. In this underground world, many things are going on simultaneously. There is a continual state of competition with many more representations and schemas activated than are ultimately used to execute tasks and resolve problems. What looks like the 'selection' from the most suitable activated representations to do the job in hand turns out to be not so much the result of calculated decision-making but rather a blind application of a law of the jungle, a state of constant competition with strongly activated representations and schemas dominating the weaker ones. Because of this, the question arises about who or what is in control down there to avoid a state of continual chaos and confusion.

DOI: 10.4324/9781003606536-3

# The Family

## Systems Forming a Network

Before we get into a more detailed account, the metro map should do its job of setting out the basic system that is the mind. The map itself could be said to represent the 'anatomy' of the mind, while its operations represent the mind's 'physiology', the way it operates in real time.

A first glance at the map presents the mind as a single, unified system. Looking closer, it is a *family* of systems. These are, in one sense, at least completely independent of one another. It is therefore a very egalitarian family: changing circumstances and current goals (conscious or subconscious) will dictate how they combine forces to solve a given task on any particular occasion. In other words, there is no central supervisor that dictates their interactions: the locus of control switches all the time and is based on the mind's current priorities at the time.

Staying with the metro system analogy for the moment, just as every metro station has its unique character and function, the mind's systems also differ from one another, in fact, much more so. Each system plays a distinct role in the mind as a whole, dealing with a particular kind of knowledge and assembling, modifying and managing it in its own unique way. We saw how the visual system manages visual knowledge. It was the one and only visual expert in the family. One particular kind of physical signal is what triggers visual processing.

## The Differences Between the Mind and a Transport System

The individual systems have already been displayed randomly located around one or the other of the two rings of the simple metro map. Thus far, at least, it bears a resemblance to a transport system. Its organisational network may not seem nearly as intricate as its equivalent in the physical brain. Nevertheless, it is not such a simple network at all if we add more detail.

One important absence from Figure 2.1 was the set of fixed connections, pathways that criss-cross between different systems

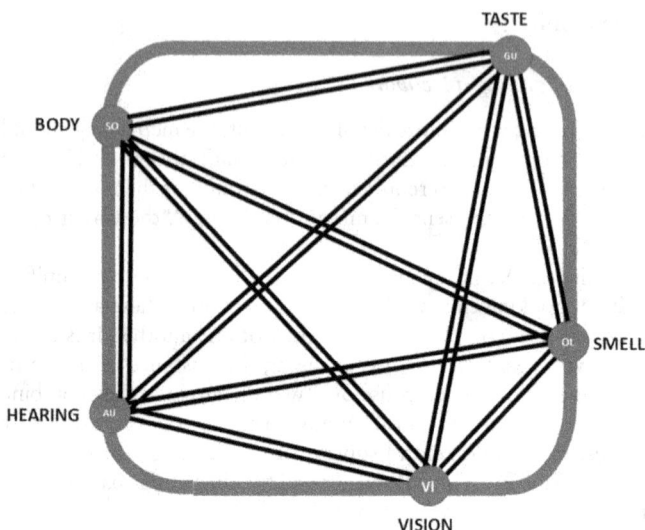

*Figure 3.1* Connecting pathways.

or, more specifically, between their *store*. The double lines shaping each ring are not pathways. For a quick impression, Figure 2.2 shows all mutual pathways between outer circle stores: in fact, these stores also have pathway connections to inner circle stores. To keep things simple in the following chapters, the connecting pathways will be displayed selectively according to the chapter's topic (Figure 3.1).

While representations themselves never move from store to store, *associations* do get formed between representations of different types. One could therefore say at least that there is some *information flow* between systems. An association running from store to store can influence what happens in the second store. Take associations formed between visual representations on the outer ring and conceptual (meaning) representations on the inner ring. Visual experiences of *white* swans will have led to an association, within the conceptual store, of two meanings 'swan' and' white'. Then the visual experience of a *black* swan will lead to the additional association of the meaning, 'black', with the meaning, 'swan'. In both cases, the original association that triggered changes in the *conceptual* store was one that was formed via the pathway running from the *visual* store.

## Control

### Who or What is in Charge?

If the mind is a knowledge network and different tasks are handled by particular combinations of these 11 knowledge systems without any one of them acting as a coordinator, the question is how what actually happens comes about. We do various things and behave and respond to situations in different ways, much of which is spontaneous and not carefully planned. Where no conscious decisions have been made, surely something else must be doing the coordination?

### The Tip of the Iceberg

As will be further discussed in later chapters, your mind's activities take place *sub*consciously and typically without us having to make deliberate decisions every second of the day. 'You' and 'your mind' are not confined to what you can reflect on, plan and generally make decisions. This would otherwise be no doubt a welcome source of control that would keep *all* of your mind running in the way 'you' want it to run. In life, this plainly and frustratingly does not happen, but given the billions of decisions required, any such frustration is misplaced. Nevertheless, it would be comforting to know that you have complete control over your thoughts and emotions.

So, to sum up, what you are consciously aware of is only the tip of the iceberg and even what you are or can become is not so easy to keep under control. We can see things that are not there. We decide *not* to do something and find ourselves doing it anyway. Since so much happens in your mind that you cannot be aware of, but which explains much of your behaviour and personality, you might indeed assume the unsettling presence of some hidden controller. What (or who) is keeping order and seems to be making those thousands of subconscious 'decisions'?

If you do a bit of mental arithmetic, for example, dividing 200 by 5, you might choose to visualise the arrangements of numbers in your mind as you do the calculation. Just consider the following: how did those images you called up actually get there? And what about the actual number 40, that just popped into your mind when you finished the calculation: where did *that* come from? The answer to all this is a little unsettling to say the least.

### It's Not All Bad

Firstly, it's certainly not the case that you have zero conscious control over anything. Secondly, there is a lot to be said for the subconscious level of processing, with its operations fully hidden from our awareness. Beneath what may be some very calm, conscious thinking or daydreaming taking place in that small tip at the top, there is, as already suggested earlier, a vast hidden jungle in the mind which is far from calm. In fact, it's quite savage with a lot of competition going on (and more of that later). Apart from the size of the hidden part compared to the visible tip, this intense activity makes it a very strange kind of iceberg. Luckily, that vast area is completely concealed from us; so, what goes on there, good or bad, generally doesn't bother us at all. If it does occasionally

bother us and we become disturbed, we often can't be certain why we feel that way because the hidden causes have to be guessed at or revealed by some expert. This was of special interest to Sigmund Freud, whose research helped to create an awareness in the general public of the subconscious and its potential for influencing us, even though his views on psychoanalysis have proved controversial. Finally, subconscious activity places fewer energy demands on the brain, whereas conscious activity, especially intense thinking, really burns up those calories.

## *A Repertoire of Coping Strategies*

Constantly changing circumstances somehow dictate most of our decision-making and our states of mind. Many choices are made subconsciously, and any conscious choices that *are* made can often be subconsciously influenced. What stops our apparently unsupervised minds from becoming completely chaotic as a result is the same principle that helps to keep us healthy and alive. This is a constant subconscious drive towards coherence and equilibrium. With unresolved problems in or between any of the mind's systems and a general lack of balance, a state of tension arises during processing and requires resolution. This means that all the time, the mind, that is to say, all its contributing systems on both rings, is doing its level best by constantly striving to make some coherent sense of what it is currently being exposed to. Importantly, this is a wholly automatic process, adapting to changing circumstances. It keeps us, as much as possible, in a safe, stable and advantageous position in order to deal with what is happening now or is about to happen. In the brain, this self-regulating process is called *homeostasis*.

## *Solutions*

All the time, every second of the day or night, our minds and brains have a great deal to cope with. We cannot possibly handle all these ongoing complexities via conscious reflection. Thinking can be slow and laborious compared to the instant responses we so often produce. Those familiar with Daniel Kahneman's book, *Thinking Fast and Slow*,[1] will know more about this. How on earth does the mind manage?

Continuing experience and instinctive behaviour provide the mind (and therefore us) with a continually developing set of strategies to help cope with future challenges. Some of these have actually been gifted to us in advance to help us survive and thrive, especially during the early months of our lives. The rest of them have to be learned. This repertoire of strategies, or the majority of them, is put together or continually adapted in response to the millions of past experiences in different circumstances. You might say that, in a certain sense, it is *circumstances* that direct our behaviour and trigger those strategies that are appropriate at the time. Or at least it is, more precisely put, what controls our minds are the circumstances *as interpreted by our minds*.

### *Captain of My Soul?*

What about those conscious plans and decisions we make, then? Well, they are not completely out of the picture. They do count as part of the decision-making that takes place. They form an upper layer of conscious responses that we can deliberately devise when we feel challenged by situations of which we are aware. This is where our hopes about possessing free will are pinned. Nevertheless, our conscious responses come from the tip, that small upper layer above a vast repertoire of hidden strategies and biases. These hidden influences constantly affect our conscious thoughts, emotions and behaviour. In other words, even our conscious decisions are not entirely free from influence from 'below'. This is the problem with those biases: they are hidden.

Our conscious decision-making, despite the fact that it gets influenced by things we are not aware of at the time, is much more like the kind of supervisory system mentioned earlier than the decision-making 'underground'. We still have to admit that *none of us* is literally, as the poet William Ernest Henley once said, a 'master of our fate' and a 'captain of our soul'. Wishful, reassuring thinking expressed in verse. Even this is still good advice for those feeling helpless; we have to go with whatever conscious control we seem to possess, sailors at the helm or on the navigation bridge, trying to steer our mental ships through the oceans of life in the smartest way we can devise while deprived of any contact with the vast engine room below.

## Spirits, Souls and Robots

### *The Biological Mind and Other Notions of the Total 'YOU'*

The ultimate motivation behind ways of describing and explaining the brain and the mind is the desire to advance our understanding of the human being that is *you*. As already stated, we will not be dealing with other potential, spiritual and metaphysical explanations. We will also not be dealing with the mind as Ryle's 'ghost in the machine', that is, as an imagined entity distinct from the body.

### *Mind Extensions*

Some like to think of the mind as something that extends beyond the minds of individuals and with, as it were, a life of its own. It is a nice idea and recalls ways of describing communities of, for instance, ants or bees or flights of birds that appear totally coordinated. If we stick to the perspective in this book, it is quite another thing to say as some have claimed that a) the biological mind as here defined, can literally expand out beyond the confines of the body except in a strictly imaginative way or b) continue to exist after the brain has died or c) come back in another body or indeed d) survive by being furnished with substitute, non-biological hardware that perfectly mimics what it replaced, thanks to some future feat of technology, all of which seems highly improbable.

In short, there are ways of thinking about the mind that permit these above-mentioned possibilities, but not in the down-to-earth version of the mind presented to you on these pages. They are imaginative, metaphorical solutions to explaining life and the mysteries surrounding human abilities, and they may have their uses, but they are without any solid foundation in contemporary science. Our 'biological-software' mind is sufficiently amazing and awe-inspiring for present purposes. For now, and for most of the following pages, it's to be 'feet firmly planted on the ground', also exploring different possibilities, but as much as is possible in tune with contemporary science. A little more will be said about these other issues in the final chapter.

## Cognition

### Cognitive Science

Talking of science, this book falls within the large and varied multidisciplinary area known as *cognitive* science or 'the cognitive sciences'. This is the range of disciplines that participate in the general study of human and animal *knowledge* (including emotions). Definitions vary from more restricted to more all-encompassing aspects of the mind. This book adopts the broader definition: cognitive science is basically the scientific exploration of mental states and processes and their manifestations in brain structure and in society. This is reflected in the way knowledge is defined in this book as information of different kinds that go well beyond the idea of knowledge that we deal with consciously and record in the form of text.

### Knowledge as Mental Representations

Adopting a pointedly broad notion of what 'knowledge' consists of means that it covers more than the everyday notion of what we say or *think* we consciously 'know' or don't know as a 'fact'. Knowledge covers not only what we like to think of as the factual knowledge which features in our conscious reflections and conversations. It also includes the vast amount of 'underground' knowledge which, apart from what we start life with, the mind has acquired and stored subconsciously. Only some of that knowledge can we, in one way or the other, raise to awareness.

### Surviving and Thriving

Some knowledge in each system is pre-stored in order to enhance our chances of survival. This is the 'inherited' knowledge with which we are equipped *at birth* and which already shows up in the very earliest stages of life, like a baby's instinctive responses to particular types of sensory experience, like face shapes, the sounds of language and the gripping reflex. Others show up later following a growth schedule like, for instance, standing upright, recognising the perspective of others than oneself and the onset of grammatical development.

*Separate but Interlinked*

Given that the physical organisation of the brain and its activities can, to some extent, already be directly observed and described, the question arises as to whether cognitive science can explain the way in which the *mind* is organised. More specifically, how can this be done in a manner that can be readily visualised and understood, at least in broad terms? As already demonstrated, knowledge systems can indeed be shown to be neatly confined to single locations on a map. Circles, lines and boxes used to show how the mind is put together do not have any physical existence as such. They can, however, be linked in an informative manner to the differently organised, intricate neural networks inside our heads. That is, of course, a major challenge for research. Exploring the relationships between knowledge in the mind and neural networks together in the brain will always remain interesting and productive in advancing our understanding of human cognition. That is precisely why it is important to develop independent accounts of the (biological) mind that can facilitate this process.

## Note

1  Kahneman (2011).

# 4 Reflections of Reality

## Introduction

This chapter considers how we create and use the knowledge that has been stored in response to ongoing life experience. How reliable a reflection is it of the reality that surrounds us? And what should we make of the inner worlds created during dreaming when the onset of an unconscious state deprives our perceptual systems of most or all of the information normally received from the environment? The restless, creative mind is still active enough to generate consciousness, but one based almost entirely on its existing internal resources without any external information flowing in and reminding us of what we regard as outside reality. This discussion will avoid most of the philosophical ramifications of what reality is and follow the logic of the adopted perspective.

## Knowledge in Practice

### Oranges and How They Seem to Us

Let's now take another concrete example of what has been discussed thus far concerning sensory perception. We'll stay with vision for the time being since we humans have a reasonably sophisticated visual system, which we regularly turn to for illustration when explaining how systems work. This system also serves as an example of how all other systems work as well.

Assume, now, that my mind has constructed a way of representing a particular fruit, say an orange. I did not know precisely how it accomplished that. I don't have a sense of the detail involved to give me even the basic idea of an orange, let alone all

DOI: 10.4324/9781003606536-4

its physical properties. Recall that included in this complex representation is a bunch of smaller representations, features that I have (subconsciously and sometimes consciously as well) learned to associate with this particular fruit. They will certainly include all the shared characteristic features of an orange, its feel, its colour, taste, texture, smell and so on. Knowledge of what an orange is for a given individual, therefore, involves different combinations of associated representations not only existing *within* one store but also extending to *other* stores.

Activating together in the form of a schema, the many component representations that make up my complex representation of an orange are something that can occur in very different situations. Activation will allow me not only to *identify* an orange whenever I see or smell one. It will also happen in the absence of an orange, whenever I *think* or *dream* about one later or when I want to *describe* an orange to someone else. Although completely unaware of all the subconscious operations involved, I will, in different situations, be able to instantly conjure up visual and other experiences of an orange in my mind. Representations originally formed on previous encounters are available in various stores and ready to be reactivated.

### Wake Up Sleepy Orange!

Take, for instance, just the *visual* representation of an orange. This is the result of a biological piece of programming that we don't have to consciously learn or bother about in everyday life. It just happens. When this particular bit of the mind's software is active, so is also the brain's visual system that supports it. The representation could be compared to an image file that gets stored in your phone or PC. If not 'woken up', no image will appear on your screen, that is to say, as a visual experience in your mind's eye. This should suggest straightaway, and correctly, that at any given moment, relevant knowledge is available in your visual store and with associations in other stores as well. This knowledge can either be in an 'active' or in a 'dormant' (inactive) state. Even when your eyes encounter a particular round shape that happens to be something very *like* an orange, the orange schema containing the visual representations and its various associations also gets 'activated' as your mind tries to identify the object. Also, when you remember or think about an orange without having one actually in front of you, everything gets activated in a similar manner. Note in passing that whereas the equivalent of an activation in a computer is a straightforward on/off matter, by contrast, activation in the mind is a matter of degree, as will be explained more fully in Chapter 6.

### Summing Up so Far

Just to recap, we have established that activating a visual representation is a bit like running an image file: it makes the object, in this case an orange, appear on the screen, although this time, the screen we are talking about is a metaphorical one, in your 'mind's eye'. This happens either when the object is in front of you or when it isn't actually present. When there is no orange in front of you and you are just imagining it, your mind summons up all the available visual representations, which together form in your mind an image of an orange. It may be a faint image or a very clear one, but it is the creation of your mind, and you normally take the rapid process of creating the experience for granted.

### The Curious Science of Optography and the Case of Fritz Angerstein

There is, nowadays, much public interest and unease about claims in the media that we can, with the aid of brain imaging and neural implants, literally read all the details of someone's thoughts and dreams. I want to sound a note of caution here. The idea of relating abstract mental concepts to physical phenomena, ones that you can examine directly, is not new. The question is, assuming that it is possible, how it can be done. Someone in the 19th century, who was concerned with ways of tracking down murderers, had an intriguing idea of what a visual representation of some perceived object consists of. In line with the claims of what was called 'optography', they thought that if you examined the eyeballs of the murderer's victim, you might find it preserved there, in patterns of pigments ('rhodopsin') on the retina, an identifiable image of the last person that victim had seen before they died: in this way, the identity of the person who had committed the crime could be swiftly revealed. The rhodopsin pattern discoverable in the eyeball will still not be the visual representation created in the head of the victim, and certainly not an image that would be interpretable by the police. It is not easy to imagine even a modern specialist looking at it and reinterpreting it as the recognisable face that could identify the murderer. Admittedly, eyeballs are linked to but not actually themselves thought of as part of the brain. However, these interpretations based on rhodopsin are reminiscent of the general desire to have straightforward translations of patterns of visible brain activity into complex thoughts and the necessary technology to make that happen.

At the time, German police investigators became open to the idea that 'optography' could become part of their standard techniques for investigating crimes of murders. A certain Fritz Angerstein was actually convicted on the basis of alleged evidence obtained from the eyes of his victim. He was subsequently executed. Fortunately, he confessed to the murder before that happened; so, on this occasion, no horrendous miscarriage of justice occurred. In point of fact, the so-called evidence could never have been held up to any serious scrutiny, and the technique was very soon dismissed as a viable forensic tool. Optography quickly went out of fashion.

## Reality as Construction

### *Does an Orange Out There Really Exist?*

Given what has been said about our experience of the world outside being based on an inner mind-based world we have created, you might sometimes wonder whether *any* experience of oranges and indeed the rest of the world outside is not just figments of our imagination. Recall that we are talking about the orange experience caused by patterns of light waves 'out there' but now registered on the retinas at the back of your eyes and only *then* being interpreted as your *personal, internal and creative* interpretations of those raw sensations. That round, orange-coloured object in all its visual detail is what your mind has made of these physical sensations picked up in your eyes. We can say the same thing about all the other sense-based representations of the outside world that lie inside us, the smells, sounds, tastes and bodily sensations. It somehow makes minds seem very individual and even isolated, with everybody carrying around their own personal version of outside reality.

Our inner world contains our relationships with events, places and people that we know and those that we don't know but that we assume exist. Only in this sense is our mind 'extended' outside us. It stretches out well beyond what we are immediately experiencing, things we have never experienced and places we have never visited, like outer space and the deepest parts of the ocean. All it needs is for us to have developed internal representations of them. At the same time, this vast expanse of reality that we sense outside us still remains firmly lodged *inside* each individual mind. This is true even though our minds 'mislead' us by providing us with powerful sense and conviction that what is inside us is actually outside in the precise form in which we perceive it. Psychologists may tell us that there are no colours in the outside world, just variations in light waves. However, most of us are still not convinced that the fine coat that we have just bought isn't intrinsically and quite independently blue for all to see. But it isn't.

### So is the World Outside Merely an Illusion?

That idea of an illusion, the philosophical reflection that you might have had, has been a seriously considered topic. Some philosophers have certainly argued about whether oranges or anything else out there *really* exists at all. Is there something out there that we call objectively real, that is to say, independently of how we experience it? A commonsense answer is immediately 'yes, of course there is. Don't be ridiculous!' Still, given what we know about the way minds construct reality, it becomes a question that is no longer ridiculous. Somewhere, people are probably still arguing about it.

Going back even a few centuries, in an account by James Boswell, the 18th-century Anglo-Irish bishop, George Berkeley, when discussing the issue with Boswell and Dr Johnson, was saying something similar. The bishop argued that matter did not exist at all in *any* form whatsoever: only God exists outside us. According to his friend, Boswell's account of the incident, Johnson expressed his rejection of this idea in a straightforward fashion by giving a nearby stone a kick. By so doing, he assumed he had provided irrefutable evidence of its independent existence. Squeezing an orange would admittedly have had the same effect and been less painful. However, in those days, when oranges were a luxury item, it would also have been an unnecessary expense. The point still is that the rock is only 'rock' as it appears in our minds. What caused the pain in our toes when striking it was something existing out there that we became aware of via sensations in our somatosensory system, its appearance conveyed by patterns of light waves impacting on our retinas.[1] However, it might feel or appear to us, you could feel right in claiming that at least its independent physical existence was confirmed by those convincing somatosensory signals radiating from a particular location on our foot, along with powerful negative representations signalling harm and originating in our affective system (the topic of Chapter 9).

In practical terms, in order to navigate our environment without incurring injury or death and to understand as humans where we are and what we are doing, we are pretty much forced to assume that stones and oranges *do* exist in exactly the way we humans are experiencing them. This is, after all, the assumed and seemingly reliable world that, for example, theories in physics describe and seek to explain. It is also a world where the behaviours of insects and animals suggest that the same obstacles exist for them as well. Although our world is always going to be inside and personal, we can usefully maintain the idea that a large part of it is identical or very similar to the internal world of other humans. We can share both simple and complex information with other people about the world outside (thanks to language). We can compare notes and argue about it. This all makes the idea of an existing world outside that we share together reassuringly plausible, as was graphically illustrated by Dr Johnson's kick.

## Ghostly Experiences and the World of Dreams

Given that all our conscious and subconscious life takes place internally and the world outside is an interpretation of individual minds, it is not surprising that people see different things when they look at the same scene in front of them. Nor is it surprising that some people, for example, see and hear things that, at the time, no one else can see and hear. Ghosts and mysterious, unaccountable 'imaginary' voices, smells and touch sensations are in many ways hardly different from 'actual' equivalents on whose existence we can all agree. If required, we can of course try and find plausible evidence by capturing them on recording and measuring devices that have been designed to tell us what we want to know.

Ghostly experiences may sometimes differ minimally from other things that are not physically present but which we can *deliberately* conjure up in our minds, using our imagination or when dreaming about them. Did we really see a ghost or did we imagine it? If two people saw the same ghost at the same time, was this a mutual creation of their imagination aided by the power of suggestion? All intriguing questions with no definitive answers.

Luckily for the most part, our minds have a way of categorising these different types of internal experience, – a reality check as it were – so that we can normally tell one type of experience from another, 'real' or 'imaginary'. However, any mind can slip up occasionally without us realising; so, it does not necessarily take drugs or illness for our minds to create illusory experiences that make us sense things that seem real but for which there seems to be no supporting evidence. It may even happen very frequently. Some, indeed, claim that we see ghosts regularly without even realising it. To know more about this subject, Oliver Sacks's fascinating book, *Hallucinations*,[2] is well worth reading.

### The Restless, Creative Mind

When we are asleep and dreaming, the brain/mind lacks immediate ongoing input from the outside. Those systems on the outer ring are not getting much, if any, information coming in through their entrance gates. The outside world is all but cut off. Then, deprived of most of what is happening outside us, our busy minds raid all their various stores in order to create alternative realities to satisfy the hunger of the mind to remain stimulated. We start dreaming. All the systems, even those on the outer ring, start to buzz. Things get activated in the stores. The realities we start to experience no longer have to conform to what is currently happening outside us, and our physical body does not need to react. We might show some external signs of responding to dream events, but they will mostly be muted versions (mumbling, tossing and turning) of how we would be reacting if those events were actually taking place in our outside world. A half-awake state[3] like lucid dreaming comes somewhere in between waking and sleeping, with some more input beginning to trickle in from outside and some querying of whether we are dreaming or not. The virtual dream realities, our *altered* state, can get temporarily merged with our dawning awareness of what our senses tell us is going on outside.

### The Content of Dreams

Self-created dream realities often seem to play out as disguised versions of worries that we have been experiencing in our waking state. Those representations that participated in our waking state are still active enough to be recalled for duty in our dreaming state. You could argue that we are in a special kind of self-created *virtual reality*, one in which we are perfectly conscious. In that dream state, we can think and have simulated sensory experiences. With temporarily reduced access to the physical stimuli coming in from outside us, the 'sleeping-awake' mind simply seeks and finds what it needs in its well-stocked knowledge stores. Following its inner drive to create coherence in all circumstances, the mind assembles a new reality freely using its current resources and in various new combinations as it constantly strives to adjust and make sense of everything coming its way.

We may conclude that the mind is never completely switched off. Patients, even when they have been fully anaesthetised, sometimes report strange experiences that they can remember and report. It is difficult not to include such experiences in this comprehensive explanation of consciousness. Despite the loss of access to ongoing events outside, the mind needs to go on creating our current reality, even in the brain's final dying moments, that is, according to recent research.[4] In Chapter 5, we will look more closely at how our inner reality in general is built up during our lifetime. This will be followed by a closely related chapter focusing on a concept crucial for knowledge creation, one which has already popped up a number of times, namely *activation*.

### Notes

1 Immanuel Kant made the distinction, roughly phrased, between *phenomena* meaning things as they appear to us and *noumena*, the things themselves, existing independently 'out there'.
2 Sacks (2012).
3 Sleep walking could be called a half-awake state because subjects are to some extent able to navigate their physical environment but in fact there is no awareness of what is going on.
4 Xu et al. (2023).

# 5   Building our Inner World

## Introduction

This chapter goes into more detail about the growth of knowledge and about how we construct our inner version of the world. For knowledge to be built, used or modified on a given occasion requires 'online' mental processing. Knowledge can be seen not just in terms of a particular representation or representations in a particular store but more generally as a combination of different types of representations that may well be distributed across different stores and variously associated with one another in the form of schemas. Calling something a 'representation' means it is a particular chunk of information (knowledge) that, in most cases, has been created or modified and stored as a result of experience. As already mentioned, this knowledge can then be activated on subsequent occasions, at the same time coactivating anything that has been associated with it. Ultimately, although much is hidden from awareness, our knowledge at any given moment is contained in the sum total of what is currently contained in all of our current schemas.

Individual representations in a store, along with any representations associated with them in schemas, were described as being either *activated* or *inactive*. Activation of one representation will spread immediately to all others currently associated with it. However, already present at birth, there are a number of ready-made representations in each store called 'primitives'. Some of them are already associated with one another; for example, in pre-formed schemas. These 'primitive' elements and their pre-formed associations do not have to be learned from experience. They provide the mind with its combined set of 'biological starter

DOI: 10.4324/9781003606536-5

packages'. Each of the 11 systems possesses its own primitives, building elements from which representations can then be constructed in their particular store. Some pre-formed schemas are capable of later modification as a result of experience.

## Representations and Processing

### Online Processing

We continue here to examine knowledge structures (representations) and how they interact with one another. Both the *use* of existing representations and any additions or modifications require online processing. The mind appears, one way or the other, to be active pretty much all of the time: this means that, at any given moment, particular representations and schemas are in a state of activation (and possibly being modified in the process) while others are resting.

### Interlinked Stores

As was shown in Chapter 4, we have various repositories of knowledge. These are the 11 individual stores that populate our minds. Stores are linked together by a fixed network of connecting pathways, sometimes also called 'interfaces'. Activation of representations in one store will coactivate any existing associated representations in neighbouring stores. It is also via the pathways between stores that new associations can be formed. During online processing, any given system's dedicated processor can create, modify and add on to its stored representations applying to its own, unique construction principles.

### Knowledge Enrichment

There will be many associations that are formed between representations *within* their own store. However, the mind's resources can also be combined by associations formed between different types of knowledge in different stores, that is, as schemas. The enriched knowledge resulting from such combinations is illustrated in Figure 5.1. Many associations involve the *inner* ring

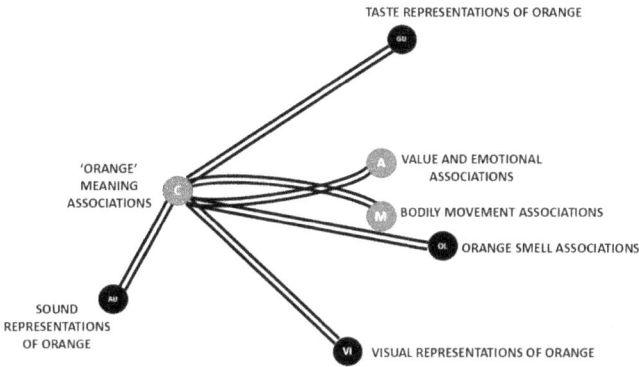

*Figure 5.1* A few orange associations.

stores. One of them is featured in this figure, namely the 'conceptual' system (C) that handles meaning representations (the topic of Chapter 8). Figure 5.1 gives a rough idea of how the various systems on the metro map would cooperate when 'orange' meanings are activated in the conceptual system. The orange schema is therefore a complex *multi-systemic* representation, and this is the norm for anything we regard as a single idea. In this particular, simplified example, the overall complexity of the 'orange' representation is actually not that impressive. In fact, such a schema will typically have a great many more associations.

## Interaction and Collaboration

When considering the simple metro map in Chapter 1, there is one final but extremely important question to raise. You might ask yourself: 'how come such a neat set of 11 systems arranged around just two rings can account for the amazing and very diverse things that the human mind is capable of?' Surely life's chaotic, changing and complex nature needs more than this. What about the work of the great composers, poets, novelists, philosophers and artists? What about the achievements of engineers, architects and great athletes? Is the mind really just a small set of knowledge boxes? The answer should become increasingly clear as we proceed through the ensuing chapters.

## Knowledge as Complex Representations

As was discussed in Chapter 4, what we think of as a single visual image (like an orange or a dog) will itself be a combination of various smaller visual features that make up the object, such as shapes and colours. These individual features in the store are representations in their own right. Associated with one another, they form a representation that can be called 'complex', every feature being composed in the code of the store in question. In this case, this will be the *visual* code, the mind's version of what a digital image file is composed of. The smallest, indivisible features in any store are its primitives that belong to its particular biological starter set. Without the visual system and the contents of its starter set, the little human starting out in life could not build the unique visual representations that distinguish human visual knowledge and human visual experiences from those of other species. Note also that the child's mind may have to create new combinations of visual features from experience, but, at the same time, always uses the primitive elements and visual construction principles provided by nature.

## Knowledge Development as Schema-Building

Representations can be complex in different ways. They may be complex with reference to their associations *within* a single specialised knowledge store, like the concept ORANGE, which is composed of items *also* in the conceptual store, like EDIBLE, JUICY, FRUIT and SOFT. These form a combination of associated concepts formulated in the conceptual code, using primitive conceptual features such as ANIMATE, HUMAN, EVENT, PLACE and ACTION. However, given the combinatorial possibilities of associations between *differently* encoded representations in different stores, the mind thereby acquires a dramatically expanded and enriched knowledge resource.

# Schemas in Action

## Schemas Awake and Schemas Asleep

Staying now with the orange example, as our life experience of oranges develops, more associations can be automatically created and added to our personal 'orange' schemas. Experiencing an orange with or without the physical presence of an actual orange is based on a particular schema being woken up, that is, being activated. Later, the relevant representations will return to a resting state. Other schemas will be or will become active, and many of the representations participating in the orange schema may well be involved in them as well: each individual representation will have its own activation history. Just as some brain activity is always present, we have to assume a constant flow of changing activation states in the mind as well. While we live and breathe, our mind/brain is never 100% inactive.

*Spreading Activation*

As already indicated, creating a representation in response to an unfamiliar experience must involve online processing activity in the mind. This is in order to carry out the construction or, given that no representations are entirely new,[1] the necessary 'modification'. If you have recognised a recently encountered smell as a *familiar* one, it means, firstly, that its olfactory representation is already present in your olfactory (smell) store: this representation was originally formed in response to what was then the new smell sensation. The re-activation of this olfactory (smell representation) might now result in spreading activation into the gustatory store and coactivate a *taste* representation: you smelled something you know the taste of. The recognition of that smell will have also coactivated a meaning in the conceptual store, for example, FAMILIAR. A fuller identification of the smell, recognising *what it is a smell of*, would activate yet more meanings. There might be additional conceptual associations such as FOOD, STEW and BURN; so, what you would then become aware of would trigger further activation, thereby generating such thoughts as 'Smells like food burning!' or 'Oh, no my stew is burning', depending on the wider context. In this way, an almost instantaneous spreading activation will take place in a continually expanding schema.

Summing up, you now have an accumulation, in the above example, of different types of knowledge representations all associated in the form of a schema. From now on, when one is activated, the other ones will also be co-activated. A normal schema in practice will feature many complex representations, and each individual representation in a schema will have and coactivate its own network of associations. Think of what might be associated with BURN, for example, pain and danger. This means much more extensive, rapid patterns of activation flooding across the mind's various stores. The resulting schema will therefore be considerably larger and more intricate than the small one just illustrated. Imagine, too, how just a smell can evoke a past memory, stirring up all kinds of further associations involving sights, sounds, meanings and emotions. Note that it doesn't matter whether or not a corresponding, comprehensive *neural* account exists: a plausible psychological account consistent with the current framework is already possible.

## Ready-made Help for Building Knowledge

### Biological Starter Sets

This now brings us back to the question of what, if anything, do we start off with in the first moments of birth (or indeed before that, in the womb)? Are these knowledge stores completely empty, waiting to be continually filled by life experience? That would be the *tabula rasa* or 'blank slate', an idea long discredited in psychology. So, alternatively, might there then be something to assist learning and *already*, in place for immediate use? The answer to this question was a definite 'yes': this is where the biological starter sets mentioned earlier come in. These biological starter sets provide the mind with 'toolkits' to get the learning process started: the primitive elements from the basis for the continual accumulation of new combinations. For example, whenever a baby learns to represent a particular sound pattern in its auditory store, the auditory representation is constructed by putting together a combination of (human) auditory primitives that are already available at birth to represent any sound that is registered in the ears.

Far from starting life on empty, each specialised store is equipped with its own set of primitive representations plus its system's dedicated processor for assembling and managing them on line. The question, *'what* is there at the start', is therefore an important one. At the moment, a full complement of answers to this question concerning the software side of the story is not yet possible. Ideas with respect to all of the mind's systems need to be developed by those who have the appropriate expertise for each and every one of them. Chapter 15 will contain a brief illustration. As to the *why* question – why are they there in the first place? – it all comes down to the way in which we have been shaped by evolution to optimise our chances of surviving and thriving on this planet.

### Genomes and Psychomes

Like all newly born members of other species, the human infant needs its own starter sets in place in order to be able to function and survive from the first few moments of life. A person's DNA will show that not everyone's starter set is 100% identical in all its fine detail. Parts of it will be influenced to some extent by each individual's experience, but the essential parts that make us human *are* shared. This is what constitutes the human *genome*. Our species has a remarkable degree of uniformity in this respect compared to our nearest relatives: these shared characteristics provide the key guide in distinguishing *homo sapiens* from other species. The precise details are only partially known about how this is manifested in brain tissue, but here, relieved of the obligation to discuss the details of DNA and account for everything in strictly neural terms, some idea of how it works can be provided here at the *mind* level of description. Ideally, the resulting characterisation of what makes a mind human will help form what might be called a parallel human 'psychome'.

### The Mind's Building Blocks

The biological starter sets consist, first and foremost, in the primitives that are specific to the system they belong to. In each store, there is a special set of primitives available for the human individual to use when encountering the kind of experience which that system is shaped to handle. Visual examples would include primitives that help to distinguish different categories of colour, shape and texture.[2] As mentioned above, some of a store's primitives will already have pre-formed associations with other primitives *outside* their store. This means that there are already some schemas involved – you might then call them 'primitive schemas', for example, the association between particular somatosensory sensations and motor representations in the motor system. They manifest themselves as various instinctive reactions like the parachute response when babies throw out that arms when reacting to a particular sudden moment to stop themselves falling. Doctors and nurses will check such responses in infants to see if everything is working properly. As already mentioned, the contents of each starter set are used by the processor of the system in question to create new combinations over the lifetime and in response to experience. In this way, the store of each system gradually gets filled with new representations of particular aspects of the experienced world.

### Human and Feline Worlds

Representations born of experience are, in different ways and in different degrees of complexity, formed over the lifetime. This results, for each individual and for each species, in a unique inner version of what lies outside us, including the bodies we inhabit. It results, for example, in uniquely human ways of visualising our environment. A cat sees the world around it in a very different way. As soon as the newly born kitten opens its eyes for the first time (after about two weeks), its visual experience will be the basis of its first new visual representations of the outside world. These will be constructed according to its own *feline* visual principles of vision, using its own feline primitives to create for itself its own feline, bluish/yellowish visual world. The differences between the feline and human visual starter sets partly very much determine this striking difference between the visual world in which cats rest, eat and hunt and the one that humans inhabit. That together with differences in the way their respective processors work. In order to pre-empt any growing sense of human superiority here, I should quickly add the reminder that the visual capabilities of eagles, dragonflies and mantis shrimps (amongst others) definitely exceed ours.

## Learning to Survive and Thrive

So far, I have been talking mostly about the knowledge that grows in each individual's head as a result of life experience. I have been assuming that it is acceptable to think of each biological starter set as a toolkit containing some sort of advanced specialised knowledge already 'pre-installed' at birth. Accordingly, we have 11 types of 'pre-knowledge' along with quite a few ready-formed associations that consequently do not need to be learned. This is another result of our biological evolution. We use them to make a rapid sense of our individual, ongoing experience of the world, especially in the early weeks, months and years, but also for learning that lasts a lifetime.

What are ready-made *associations* with other primitives, both within and across different stores, are often referred to using the term 'instinctive'. They are easily identified in observable infant behaviour: the baby orients its gaze towards faces, throws its arms if it feels it is about to fall and all those responses that are checked by health professionals. There are other less obvious examples that must be there logically in order to explain behaviour, such as the creative and complex use of language that emerges so early, so rapidly and so effortlessly in immature minds.[3]

Summing up, primitives involve more than a set of singler building elements; there are also ready-made associations. Associations can exist *inside* each store. Certain tastes in our gustatory store on the outer ring might be composed of one or a combination of taste features, bitter and sour, say. A given taste might cause a baby to reject food as potentially poisonous. The pre-formed association has evolved to protect the child; quinine is just one example as it usually triggers an instinctive disgust response.

The disgust response to a bad taste is one example of a primitive schema. The taste in question has a pre-formed association with a negative *value*. As will be discussed later, values are located elsewhere, in the affective store (A). A taste representation will also be associated with, for example, particular representations in the motor system (M) that guide physical movement (creating the outwardly observable, bodily expression of disgust) Representations in the *spatial* system (SP) will also be involved since representations of the spatial environment are needed for any appropriate reaction to be enacted. In this way, these primitive schemas permit rapid, complex responses engaging more than one system at the same time. Calling them 'instinctive' is just another way of saying they have evolved over very long periods of time and are therefore not constructed from scratch.

## Individual Differences

### *Do We All Begin Life Identically 'Pre-programmed'?*

To state that all human individuals are equipped from birth with the same starter sets in each of their 11 systems might lend support to the false impression that we are all turned out like identical, factory-produced machines. This would go against our general understanding that evolution proceeds by natural selection and random fluctuations. In one sense, it is correct to say there must be a degree of uniformity between individuals. We all belong to the *Homo sapiens* species after all. However, despite a high degree of uniformity, there is still room for variation in how these shared characteristics are actually manifested in individuals.

To reiterate the point, despite none of us having the same experience or the same accumulated knowledge, we all share *some version of* the same human starter sets. Adding this qualification (in italics) is very important since no one individual is the same, even 'identical' twins. Even the contents of the starter sets are not identical, that is to say, in every respect. Each individual, as it were, will have the same basic set of primitives at birth and have most, if not all the same ready-made 'instinctive response' schemas. However, since there is still room for quite a lot of variation, we must assume that the contents of starter sets can admit different 'settings'. After all, we are all born with our own particular personalities, and this has to be explained at the mind level as well as at the brain level of description. Settings are a useful way of conceptualising how two individuals with the same basic set of inherited (human) characteristics can still differ in their respective character traits and not necessarily identical predispositions.

### Different Software Settings

A software 'settings' metaphor is one way of explaining how individuals can differ in what could be called the range of options with regard to the otherwise shared components of starter sets. This includes differences manifested in the preformed *schemas*. Such differences have for some time now occupied psychologists and psychiatrists concerned with categorising different types of individuals, with the focus often on those that are to some degree socially dysfunctional and needing targeted assistance. There is what seems like an increasingly narrow range of individuals who fall within a category that people prefer to call 'neurotypical' (or, as I would personally prefer, '*psych*otypical').[4] Beyond this non-divergent range, measures like the DSM (*Diagnostic and Statistical Manual of Mental Disorders*) and the ICD (*International Classification of Diseases*) seek to classify people with the specific aim of identifying those who can be regarded as being in need of treatment and also those who simply differ in terms of their other basic personality profiles without requiring medical intervention. Unsurprisingly, with continuing research and cultural shifts in attitude, each successive version of measures like the DSM is never exactly the same as the previous one.

The general idea is, then, that people born with different personalities will have different ways in which their biological starter sets are configured. This reflects their 'predispositions'. Personality traits offer an easy way to illustrate this. Someone who is identified as an especially calm, unflappable person, one who is very seldom anxious will be sharply distinguished from another person who is particularly prone to anxiety, their stress levels rising automatically with even minor provocations. Both will share the same human-specific primitives, but they will differ in how these and their neural equivalents are organised at birth. Possessing the same basic set of characteristics is no obstacle to individuals exhibiting *enormous diversity*. The behavioural habits and frames of mind that are determined by an individual's starter set can undergo considerable modification as a result of life experience. They are, after all, only your 'starting' point, even if modifying them or suppressing them completely can vary a great deal in difficulty as regards how easy or possible changing them can be.

### Modifying Instinctive Responses

As was implied above, many associations that make up a complex instinctive response are typically difficult or impossible to unlearn. This is because they have turned out over evolutionary time to be of continuing value throughout life for maintaining our body and optimising chances of survival. However, since adaptability is also extremely important, certain pre-formed responses can indeed be altered by experience.

The settings analogy would work for other species. Those cats that leap into the air on suddenly seeing a snake-shaped object, whether or not it is a *real* snake, must already have in their visual store a particular complex shape representation that combines one or more curved shapes arranged in a particular sequence. This will be part of their (feline) visual starter set. They are not literally pre-programmed to recognise and fear snakes themselves, but rather particular shape configurations that have over evolutionary time proved to harbour extreme danger. The 'snakelike' visual representation will also have an association with given representations in the *spatial* system on the inner ring, since, as will be touched on later, visual patterns will need to be interpreted within 3D space.

As part of ready-made threat-avoidance schemas, certain representations in the *motor* system responsible for generating the relevant physical movements associated with escaping the threat will also be activated. Certain visual features may eventually turn out *not* to be dangerous after all, leading to a modified version of the dangerous shape: threat avoidance schemas in this way become progressively refined so that only shapes that happen to be very close to those of real snakes that occur in this cat's environment will end up triggering a threat-avoidance response. Perhaps, in time, *all* shapes that are not actual snakes will be ignored. In this way, the snake-wary cat gets better and better at distinguishing real snake shapes from harmless snake(-like) shapes. The instinctive reflex may remain but as a much more refined version of the original one. The same goes for humans but (perhaps) with other shapes, like spider-related shapes and mouse-related shapes, for example, associated movement patterns will also be involved.

Now it is time to turn to the very important topic of activation and look at it in greater detail. It will help to explain some very basic things about our minds, including *attention* and various degrees of *conscious awareness* and about how we think and how we feel. Again, we are only at the beginning of understanding these phenomena, but there is already a large research literature replete with theories and hypotheses about how to explain them.

## Notes

1 In particular, given the use of primitives.
2 Although research has identified neurons in the brain that are sensitive to, for example, edge detection or colour, the equivalent primitives in the mind need to be defined as well.
3 This is the topic of Chapter 15.
4 The use of these terms reflect the need to avoid the negative implications of calling someone as not 'normal'.

# 6   Activating Knowledge

## Introduction

This chapter looks more closely at activation: a fundamental principle governing the mind's operations. Whenever a given representation or schema becomes active, its activation is always a matter of degree. Activated representations become candidates in a competition as possible alternatives for use in the execution of some tasks. The more intensely activated candidates will have a competitive advantage over their current rivals. Resting levels are also a matter of degree. Once a representation has been activated, it will later fall back but to a slightly higher resting level than the one from which it rose, making it more competitive the next time around. Frequent and regular activation will see its resting level grow higher and higher, each time giving it a more advantageous position from which to rise when it gets activated the next time around. While inactive, a representation's resting levels will gradually decline: if representations are left inactive for extended periods, the resulting decline in their competitiveness will become noticeable.

## Activation and the Law of the Jungle

The idea of knowledge being something that is 'activated' in the mind has already cropped up in earlier chapters and is a familiar theme in cognitive psychology. It is an idea that is easily associated with the brain, with neurons firing and producing colourful patterns on a scanner. This chapter looks more closely at activation in the mind. Like association, activation is a fundamental processing principle that applies to *all* of the mind's systems.

DOI: 10.4324/9781003606536-6

The fact that mental representations are usually associated with and will coactivate many other representations also requires a discussion of how such associations happen in the first place. And what happens next? Furthermore, as will soon become clear, activation is also related to another basic principle, namely *competition*.

As already discussed, there are different encoded types of knowledge. Sitting dormant in the various stores, knowledge representations are not much use unless we, or rather our minds, can *use* them. How then are they woken up selectively in order to help solve all the tasks that the mind is continually presented with? Tasks include both those that will be carried out subconsciously and also tasks that are conceived and to some degree managed consciously, in our thoughts. We are talking specifically about how knowledge is brought into action, that is, 'activated'. It will be reflected, in neural terms, by impulses being sent over the brain's own myriad pathways, electro-chemical patterns of cells firing some within local regions of the brain and others across more distant regions as well.

As we have seen, any kind of knowledge will come in the form of representations that are combined in various ways. When you smell something familiar, particular *meaning* representations (bits of conceptual knowledge) are activated alongside whatever has been associated with them, in particular *smell* representations (bits of olfactory knowledge) in the olfactory system. In this way and for that individual, the 'meaningless' sensory perception of a smell becomes meaningful by being associated and instantly coactivated with its current conceptual representation.

### Different States of Activation

In principle, the default state for any representation may be described as 'dormant', 'inactive' or 'resting'. In practice, for most of the time in our characteristically restless minds, this is more likely to mean 'either 100% inactive or at least very close to this resting state': however, 'resting' will do as an adequate description. Becoming 'active' means that an *in*active (resting, dormant) representation has temporarily left its current resting place and is now rising up and becoming available for potential participation in a current processing operation. Another thing needs to be said about resting states, and that is that any period of inactivity is characterised by a decline in the representation's resting level. The immediate decline will be minimal. It will continue at a snail's pace until a long enough period of disuse has elapsed for its effects to become noticeable. This is because lower resting levels also mean a decline in competitiveness: the representation has relatively further to rise when activated.

Given that mental activity will be reflected in what happens in the brain, the mind, from an energy conservation point of view, appears not to be as economical as it could be. Many more representations than will ultimately contribute to the execution of some task will be woken up from their imaginary slumber. Although only some of these activated 'candidates' may still prove useful in the end, for a brief moment, they are there, temporarily activated, as *potential* candidates for the particular job in hand. Are they then 'unnecessarily' using up space and/or calories? Is the biological software making unnecessary demands on the hardware?

If all currently activated representations enter a *competition* ending in winners and losers, the winners should be those best suited on that occasion to deal with a specific task or set of tasks that is facing the mind at that particular moment in time. As will be explained further, these competitions are pretty much trials of strength, but it seems that the mind can mostly handle this jungle-like state of affairs without any problem at all. And normally, without bothering us. The fact that many potential candidate representations will turn out *not* to participate in the ultimate execution of some tasks may indeed make their activation seem like a waste of time and energy. However, the perceived circumstances may still shift, even minimally at any moment, life being quite unpredictable. If that does happen, those less strongly activated representations, while still activated to some degree, might then prove to be useful and appropriate after all. They can then be called into action more quickly from their current state of activation and relative readiness. The apparently wasteful over-activation of representations may not, in the longer term, be so very wasteful after all. Moreover, it is not even clear, in brain terms, that activation necessarily or always involves energy consumption.[1]

*Activation is a Matter of Degree*

As already indicated, activation is not a simple on/off matter – 'active' or 'not active' – but a matter of *degree*. For example, assume that experience has made us familiar with oranges. When we see an orange on a particular occasion, responding to light wave patterns just been registered in our eyes, relevant representations in the visual store will be immediately activated. Various other types of representation will be co-activated at the same time, including the conceptual ones carrying associated meanings. For example, the *taste* (gustatory) representations that will have also been formed when tasting oranges will be activated as well. However, in the current context, these may not be as *strongly* activated as the visual ones: circumstances may make taste associations less relevant. However, if we also happen to be hungry at the time and enjoy the taste of oranges, the current activation level of those taste representations will be influenced accordingly: they will become more strongly activated. Note that the visual representation triggered by the sight of orange-like shapes will also trigger *other* possible candidates that are not oranges but share many of the same or similar visual properties which have been activated. In this case, the alternative visual interpretations of the same sound wave pattern will trigger the coactivation of the candidates' own associations. The mind's global response will be in line with its general principle of maintaining coherence. It will end up with what an external observer might qualify as 'the best-fit schema in the circumstances'. In this case, it will be the orange schema. This should give a rough idea of the frantic, high-speed mental activity going on when, as far as you are concerned, you just see an orange.

## Conceptualising Activation

*Handy Metaphors*

Ways of portraying activation in visual terms may help to understand how it works and underline how important it is. The degree of activation that a representation may undergo can be shown in different ways. The preferred way here has been to use the metaphor of *height*. The degree of activation can then be imagined as ranging from zero, then relatively weak, and rising on upwards to its most intense level as displayed on a traditional thermometer or pressure gauge.

Another way of displaying degrees of activity is by using shades, starting with pale grey (inactive or virtually inactive). Then, activation can be portrayed as various darker shades, with the darkest shade indicating 'intensely activated'. This use of shades of grey or colour might suggest not height but *heat*. Deep red (or in non-colour versions, the very darkest shade of grey) would then be the hottest, most intense degree of activation. Paler greys (or reds) would be the opposite, indicating cooler, lower levels extending down to zero activation. Both metaphors can, of course, be combined. Figure 6.1 uses darker shades of grey (or red in colour versions) to indicate intensity of activation. Note in passing that consciousness (awareness) will be involved only at more intense levels of activation (see Chapter 12).

### Co-activation

It should be clear by now that once any representation is active, it will automatically cause any representation with which it is associated, either inside its own store or externally, that is, outside its

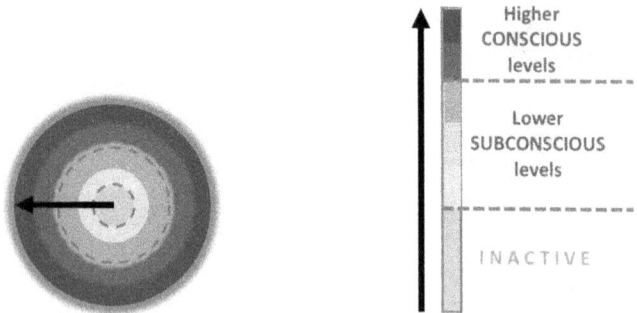

## Degrees of activation

*Figure 6.1* Two possible ways of imagining degrees of activation.

store, to be coactivated as well. The very first coactivation of two representations is what brings about an association in the first place, and this ensures further coactivation the next time around.

## Competition

When a number of representations become active, it then becomes important to know to what *level* they have been activated. The mind will activate many, many representations on a given occasion, like generals mobilising all their troops when only some of them will actually get to take part in a given operation. As already mentioned, a whole number of closely related representations get activated, with only some of the candidates eventually getting 'selected' for whatever happens to be the task in hand. However, in the mind as a whole, there is in fact no 'selector', no 'general' running the show. The current context – as interpreted by the mind – will mainly determine relative activation levels at any given moment, and contexts may shift slowly or rapidly.

## Handy but Risky Metaphors Based on Conscious Processing

Since there is no one, or no one *thing*, which is remotely like a leader that makes all the 'decisions', we have to be careful about the words we use to describe what is going on, especially when talking about all the activity going on at levels that are insufficient to involve consciousness. The commonly used term, 'selection', for example, can only be a convenient metaphor for a process that does not actually involve a selecting agent. This process would arise from a number of forces interacting to generate the outcome; so, in reality, what is viewed from externally might look like the result of someone's or something's 'decision' is ultimately the internal outcome of blind competition between candidates. Some conscious deciding may well be going on at a higher level. Effectively, what will actually happen is dictated by the internally represented context at the time. The currently perceived context will bias the result one way or the other. Conscious decisions are certainly not excluded from this process but may never completely dictate the outcome completely. Many candidate representations will be activated to take part in the current, best-fit schema to carry out the task at hand. In whatever the current context happens to be, those which turn out to have the highest (darkest, reddest and warmest) level of activation at that particular moment will have the best chance of becoming the winners.

The point here is really to emphasise how much is going on in our minds *below* the level of consciousness. It's about how the mind is running in a way entirely different from how we imagine it works when we make conscious decisions to do something or *not* to do something (see Chapter 12). Conscious decisions are not going to be totally under our control either, and so not entirely free from bias. They are typically influenced to a greater or lesser degree by those *sub*conscious processes over which we may have little or no control. Finally, the hidden competition going on between rival candidates is not a random affair. The context in which that competition is taking place is crucial because it *does* influence the outcomes.

### The Pros and Cons of Subconscious Processing

Following on from what has just been said, we can never be directly aware of anything except, sometimes, the *outcomes* of mental processing activity. We cannot be aware of the many processing details, for example, the contents of stores, what has been activated, and the degrees of activation. That all of this internal battling happens safely tucked away in the depths of our subconscious is just as well; otherwise, we would surely be suffering unimaginable and unremitting torture. Still, we can sometimes get some indirect sense of what must be going on. One example is when we are in the process of trying to recall a name, and it just won't come to the surface. It is on the tip of our tongue, but it just won't reveal itself. This can be very frustrating. The attempt to recall the word feels increasingly effortful: we know it must be 'there' somewhere, but we just can't quite get a hold of it. Much mental processing, however, seems to require little or no effort. For example, we can chat away effortlessly with a friend, speaking fast and feeling nothing of the complex cognitive processes going on to make this happen, assembling a stream of messages as well as working the muscles in the lungs and vocal tract to producing these messages as sound waves, that is, until suddenly our brows wrinkle and unexpectedly we come across something problematic like that name that we cannot immediately recall. Until that point, we remain happily unaware of all the conflicts going on inside us. It can sometimes be very useful to know what in our subconscious is influencing us, but overall, the 'pros' of subconscious processing much outweigh the 'cons'.

## Knowledge and Performance

### Different Broad Categories of Knowledge

At its most fundamental level, all knowledge boils down to the same thing: *representations* and their *schemas*. The representations all follow the same general processing principles as regards how they get activated, how they form and coactivate associations both within and reaching beyond their own store. Apart from what we already have available at birth, all other knowledge is formed in response to experience. This process is generally referred to as 'learning'.

In the psychology literature, various categories of knowledge and memory have been proposed and have become widespread. These now need to be interpreted within the perspective adopted in this book, especially with regard to activation. They usually relate to the kind of knowledge that can feature during conscious mental activity. The question here is in what sense do these established psychological categories provide insights into how knowledge is formed and activated. A case in point is the long-established distinction in the psychological research literature that is the one that distinguishes *declarative* knowledge from *procedural* knowledge. The same distinction is sometimes used to describe memory.

### Declarative Knowledge as an External Phenomenon

Roughly speaking, 'declarative' knowledge refers to the outcomes of thought processes, information that you can consciously reflect on, query, test, debate and claim to be true or untrue, like the information you are getting from reading these pages. Broadly equivalent labels include *explicit* and *metacognitive* knowledge, although each label tends to come with a particular theoretical position and focus of interest. The basis for this knowledge, whichever label is chosen, resides in the *conceptual* store. It is the conceptual system that plays the crucial role. It sits at the crossroads of different schemas that are intensely activated during 'conscious' thought processing.

As pointed out earlier, conceptual representations in the conceptual store, along with whatever they happen to be associated with, remain the same whether they are activated intensely enough to participate in thinking or whether they are activated *below* the levels required for any kind of awareness to occur. Declarative knowledge may be *acquired* consciously, where high activation levels are required. After that, it can also be activated at lower levels of activation as well. Many of the facts you have learned about a person or place may be activated on a given occasion and may influence the way you behave, but you may not always be conscious of all these facts at the time.

### Skill

'Procedural' knowledge is sometimes described as 'knowledge how'. In terms of this book, it is about representations that are regularly and efficiently activated in the execution of some skills that are not or are no longer dependent on conscious control. Practitioners of a particular skill who can perform their skill unthinkingly can, for example, direct all their conscious thinking to strategic applications of their skill, given particular contexts and particular goals in mind. The skilful execution of tasks that once appeared to rely on conscious control is sometimes described as using knowledge that has become 'converted' into procedural knowledge. However, a better explanation is probably that efficient routines have developed *separately* by themselves, below the level of consciousness and have made redundant the more ponderous, conscious attempts to perform the tasks associated with the skill in question.

## Resolving Competition

### Patterns of Dominance

Recall that activation will always involve competition between candidate representations. This is inevitable simply because the mind is so massively interconnected: one representation may have many associations. This means, in turn, that there will be spreading activation across multiple stores even though the representations that get coactivated as a result will have very differing degrees of relevance for tasks currently being tackled. At any given moment during the performance of some task, certain individual representations and schemas will dominate. In both cases, 'dominating' or 'dominant' here means 'currently most strongly activated' and hence most likely to be involved in the resolution of the task. What actually gets to dominate on a given occasion will be determined by a) how strongly established the representations happen to be at the time in their respective stores – effectively how high their *resting* levels were before they were activated and b) the particular *context* in which processing is currently taking place, favouring some representations and schemas over others.

Contexts can suddenly shift, and this need not refer to changes in *external* circumstances. A sudden memory or change in mood (like the loss of patience) can have the same effect. What is valued in a given context at one point in time, as reflected in its superior activation levels, can quickly change, and the current patterns of dominance will readjust accordingly. Individual differences also play a role. How values (in the affective store) are associated is based on the individual's own processing of what is currently going on in the external environment and on the internal state they happen to be in at the time. The main point is that what the individual happens to personally value (subconsciously and consciously), in any given circumstances, will influence what gets to dominate their minds: it will subsequently affect their personal attitude and behaviour at the time.

## Knowledge Change Depends on Activation

### Gains for Losers, Losses for Winners

Even in minimal ways, both activation *and* non-activation will affect the current state of knowledge. This is because just as resting levels decline as a result of *non*-activation, they always *rise*, however minimally, whenever they *have* been activated. Activation has this effect, to a greater or lesser extent, for any representation. Even the 'losers' in a competition experience some gain before they begin their return to a resting state. Then again, strongly established representations with high resting levels and consequently increased chances of being 'winners' can still suffer reversals of fortune. A prolonged lack of activation will then cause their resting levels to significantly decline.

### We are Constantly Learning

From what was said earlier in this chapter, you may assume that newly formed representations and associations in a schema will typically be stored at a very low resting level. For a time, it will need a big extra boost to allow them to rise up when activated and compete with stronger rivals that get activated at the same time, but from a higher, more competitive resting level. This boost may happen briefly while the representations have only recently been formed; in other words, when what is newly learned has some kind of novelty value. The standard course of events will be that the newcomers will need to undergo *regular and frequent activation* for them to finally acquire a high-enough resting level from which they can reliably rise up and prove to be competitive on a regular basis. A common metaphor used in this context is 'accessible': representations become more accessible as a result of frequent activation. We could also say they become more readily available for use by their processor. Since the activation of representations is happening all the time, we can assume that our minds are constantly gaining, updating and reviewing knowledge in some store or other on the inner or outer rings. And, again, this is happening irrespective of whether we have any conscious intention to learn.

### Competition: The Tip-of-the-Tongue Example

As was indicated above, the knowledge you actually use at any given moment is the outcome of (mostly) hidden competition. You can get a flavour of this when you consider the tip-of-the-tongue phenomenon mentioned earlier: this is when your mind is trying to retrieve a word that you are absolutely certain you know, but something seems to be blocking it. Alternative words may pop up instead, but you know they are not the one you intend. As researchers such as John Laver and others have shown, slips of the tongue can give us a sense of this hidden competition.[2] The word you are looking for is not completely blocked, but at the same time, a rival is still competing strongly with it. One popular example is the phrase, 'not in the sleast', where two competing candidates to express the intended meaning were 'not in the slightest' and 'not in the least'. When the expression was actually uttered, the competition between the two candidates had not been resolved, ending as equal winners and resulting in the (unintended) hybrid form 'sleast'. In such situations, there is no inner reflection: 'which of these two shall I use?' They just pop up as neck-and-neck rivals as different options to express a single meaning. If you had managed at the time to notice what you had just said, you would immediately have spotted it as a possible but non-existent word. This situation gives us a brief window looking in on the whole activation story, what must be going on in our minds 'beneath the surface' as well as any conscious reflections we might have at the time.

## Multitasking

Finally, I'd like to reiterate the point about the value of activating many candidate representations at the same time. With the sophistication of its architecture and the enormous computing power it possesses, the mind can do it this way often without any sense of effort or any expenditure of calories on the part of the physical brain. In other words, it is not as wasteful as it seems, even given the fact that there will always, we assume, be a limit on how much energy can be consumed by the brain at any one time. Also, the mind is a parallel processor *par excellence*, with its various systems all working at the subconscious level simultaneously on tasks only they can do. They do not have to perform slowly, one-by-one, in a sequence of tasks with one system doing its job and then another. The fact that many candidate schemas are activated together as a result makes for a highly flexible mind. A highly activated representation in one of the stores involved in executing a given task may not fit well with other coactivated representations. We have minds that can rapidly adapt to changing internal or external circumstances. As the Greek philosopher Heraclitus said, we never step into the same river twice. Events are constantly in flux, changing all the time and our minds have to match their capricious twists and turns, however minute they may be. Options that were not so relevant at one moment can then suddenly become relevant the next moment and, while falling gradually back to their resting levels but still in an activated state, they are consequently more easily accessible if needed than if they were already back at their respective resting levels. Put another way, *the mind keeps its options open*. It is also the case, as will be discussed in greater detail in Chapters 12 and 13, that slow conscious processing does require intense levels of activation and, hence, it is presumed, relatively more energy than all that subconscious processing going on.

## Notes

1 There is in fact a 'activity-silent' model of working memory (in the brain), whereby it can be maintained without any neurons firing and presumably no energy consumption. See Spaak and Wolff (2025).
2 Boomer and Laver (1968).

# 7 Remembering and Forgetting

## Introduction

This chapter interprets, within the current perspective, some familiar terms in psychology and also neuroscience with which readers may be familiar. They are related to ways in which memory and forgetting have also been a part of everyday, non-scientific discourse. Examples include long-term memory, working memory, false memories and episodic memory. In this book, 'long-term' memory must be interpreted as referring to 'all stored representations', whether they happen to be active or 'dormant'. Particularly, important is the precisely definable concept of *working* memory. This refers only to those representations and schemas that happen to be in a current state of activation and irrespective of their degree of activation. When active, the mind is constantly busy with more than one, and usually a multitude of tasks at the same time. Psychologists have debated the question of how much can be in working memory at any one time. Memories can also seem lost for good, but still on occasions resurface. Old age typically appears to be characterised by a general resurfacing of old memories, and a possible explanation for this is given.

## Describing Memory

### Storage and Retrieval

As so often, the language we use to describe various concepts reveals how we view them. Terms closely associated with memories like 'remember', 'recall', 'reminisce', 'forget' and 'call to mind' can take on different meanings in different contexts. We will

DOI: 10.4324/9781003606536-7

continue with the extended meaning of the term 'memory' to include both storage and activation. Then again, technical terms in current use like 'long term memory', 'short term memory' and 'working memory' may have their uses in a particular perspective on the nature of memory. In Chapter 6, the distinction was based on activation. Working memory was knowledge (representations and schemas) in an activated state. This activation-based definition of working memory obviates the need to make different memory categories that imply separate types of storage. In the case of 'long term' memory, this now refers to precisely the same representations in precisely the same locations (store) as they were when activated. The only relevant distinction is what is and what is not in a state of activation. You can, of course, still talk, in relative terms, about how long representations and schemas must be maintained over a given period of time to qualify for the label 'long': it may last millisecond years. Defining 'long' or 'short term' precisely will always be a more or less arbitrary matter. For example, 'short' term might suggest the period in which an activated representation begins its progress of decline without yet arriving at its resting levels. Highly activated representations should remain relatively accessible for a somewhat longer period than others.

Words like 'remember' and 'recall' suggest *activity* in the mind on a particular occasion. Other expressions like 'he has a mind like a sieve' (he keeps forgetting things) or 'he has an amazing memory' seem to refer more to memory as an *ability* – or lack of it – and as a *resource* that can be drawn on. If we take the act of remembering on a particular occasion, this refers both to the *deliberate* use and also to the *spontaneous*, that is, unsolicited activation of stored knowledge. You have either decided to try and remember something or a memory just pops up into your head for some reason you don't understand: in fact, this will be due to some contextual cue, an association with something else that you cannot identify at the time. Either way, it will often involve the notion of past time, which is a conceptual representation (past time), and perhaps even a primitive. Memory schemas will include some time stamp on them, even just a non-specific notion indicating 'sometime in the past but I can't remember when'. Whether the remembering is deliberate or whether it is unsolicited and somehow 'just happens', the kind of memory we are talking about here always involves *conscious* experience.

Remembering provides us with clear evidence that something in our heads has been stored and can, in principle, be retrieved. Psychologists sometimes make typological distinctions between different types of memory, such as 'episodic memory – the memory of specific events – and more specifically, 'autobiographical' memory – those events linked closely to an individual's life. It is necessary to think in terms of what types of schemas would correspond to a given memory type: in other words, which particular stores and representational types would be involved. Different memory types might be distinguishable when looking at how their schemas involved are manifested in brain tissue. In any case, their psychological description can all be done in terms of the general and specific principles of mental processing discussed in this book.

### Decisions, Decisions, Decisions...and Bias

Most activation by far, as already mentioned, operates without implicating the higher levels of consciousness associated with thinking. A particularly striking example of the subconscious workings of memory is when our mind seems to be making 'decisions' for us, or put another way, decisions that we are taking that we don't actually know about at the time. In other words, while we are putting together an assessment of someone or something *consciously*, at the same time, *sub*conscious processes are working in a way that can bias that conscious assessment. For example, people on an appointments committee interviewing someone for a job throughout the interview may well consciously feel that they are faithfully following an agreed, rational and fair assessment process. However, despite their best intentions, they may still be making up their minds about that candidate much earlier: what happens later in the interview is unlikely to change this subconsciously influenced, intuitive' evaluation unless there is clear evidence to counteract it. Some interviewers may be well aware of this at the time: 'I just knew she was the right person for us in the first five minutes!' Whether or not they are aware of it, in effect, the decision may be prematurely fixed before the end of the interview, and any panel discussion that follows will not change it. Discussions of ways to combat bias, for example, in the human resources literature, far outnumber empirical studies of the phenomenon, although its existence is widely accepted.[1]

Often, biases reflect general patterns within the community, showing, for example, a subconscious preference for recruiting men for engineering jobs.[2]

Intuitions may be secretly guiding opinions, but they do not have to be wrong. As the Nobel prize-winning psychologist, Daniel Kahneman, claimed, much decision-making happens not after a careful, rational process has taken place but rapidly on the basis of hidden cognitive biases.[3] These biases will have been established already for various reasons, and part of that process will certainly include the activation of combined memories of people and situations that will be relevant and will subtly affect the interviewer's judgement. If you are the interviewer, without you being aware of it, the candidate might trigger a memory of someone you didn't like and bias your judgement. If, on the other hand, you become conscious of that resemblance, you will have an opportunity to consciously control the effect of that bias.

### We Want to be in Control

Knowing all of this now is, of course, unsettling for most of us. After all, we want to be in conscious control of what we decide or do not decide. In the appointment interview example, we may indeed remain confident that we have followed the fair procedure to the very end of the interview. On other occasions, we may become aware that although we had had the firm *intention* of doing one thing, we somehow ended up doing something else that we didn't want to do. At least, being conscious and having to control all of the otherwise subconscious work would slow our mental activity down to a pace that even the slowest sloth would find slothful.[4] Apart from the little matter of driving us completely insane.

Summing up so far, when we refer to our memory in general, we are referring to a perceived ability to construct, retain and activate knowledge representations of various kinds and their associations. We have many everyday ways of referring to memory and acts of remembering. For example, we say things like 'she has an amazing memory', or 'I have a terrible memory', or 'he has great memory for details' or 'I have a terrible memory for people's names'. All these expressions can be translated into the terms of this book as statements about representation, storage and activation with the conceptual (meaning) store and its multiple associations with the other stores playing a central role. At the same time, having said that the act of remembering must involve conscious processing, it is of course also true that psychologists and psychoanalysts do talk of 'subconscious memories'.

In one sense, at least, all memories are subconscious in the manner in which they are stored and activated: they are permanently hidden with only their content, not necessarily open to conscious inspection. As will be discussed in detail later, the content of memories comes to awareness indirectly, that is, translated into perceptual form. 'Subconscious memories' clearly refer to those schemas that are activated to a level that is sufficient to influence our thoughts and behaviour but still not enough for us to become actually *aware* of their influence, as was illustrated in the above example of the biased interviewer. Psychologists, when using this term, often refer to memories of very unpleasant events and people, things that an individual's mind has kept hidden or 'repressed'. This interest dates back to Sigmund Freud's best-known work[5], which focused on precisely this phenomenon and the psychotherapy needed to bring patients relief from various kinds of traumatic childhood memories. Summing up, the meaning of 'memory' as the term is commonly used is not such a simple, straightforward one as we might imagine.

## Possible Limits to Working Memory Capacity

The question has often been raised regarding possible working memory limitations. This question could be rephrased as: 'How much activation can be maintained at any one time?' It may be, for example, that for most tasks, there is *no limit in principle* as to how much can be activated in the various stores: in that case, only processing factors (efficiency, avoidance of excessive levels of competition and general working memory clutter) may limit how much can be activated at the same time. From an activation perspective, anyway, resource limitations as regards the hardware (the brain) may start to kick in if the overall management of some tasks begins to trigger levels of activation that are very intense. In effect, this would apply to tasks that involve *conscious* processing. Apart from that, we must assume that the mind's biological hardware must have the energy resources at any one time to support the requirements of its software. Consequently, all other things being equal, subconscious activity, which typically involves simultaneous activity in so many parts of the mind (and indeed brain), must be well within these resources. Even casual thinking, where it is not particularly focused, seems to ring no energy alarm bells. Until we begin to concentrate, for a prolonged period of time, or even very shorter periods, on something which proves particularly challenging if we experience warning signs in the form of a sense of effortfulness and fatigue. The conclusion for now is that working memory, that is, mental activity in one or more stores, might in principle be limitless in those circumstances that do not require intense levels of activation. Typical tests of working memory capacity that involve the *conscious* execution of some experimental tasks will obviously not test this idea properly.

## The Suppression of Memories

When engaging with traumas and similar distressed states of mind, psychologists often talk of memories being repressed (dissociative amnesia). Events such as life-threatening moments in wartime or physical and emotional abuse in childhood may be so horrific that the mind cannot cope with reliving again and again the extreme stress it caused at the time.[6] As discussed earlier, schemas are the forms that memories take in the mind. Whereas those schemas that have been inactive for a long period will eventually undergo a substantial decline in their resting levels and hence in their accessibility, memories of certain events may undergo a much swifter process of decline. This is because the individual finds the memory of them intolerable. These memories will involve traumatic or extremely unpleasant events that have caused great distress. These arguments about the pros and cons of engaging in therapeutic interventions that stimulate the recall of such events and hence break through the suppression process. In any case, what is behind the whole idea of rapid memory suppression needs a somewhat more detailed explanation. This is the result of the mind effectively placing a protective 'taboo' on the schema responsible for representing the distressing event. The schema(as) representing the distressing situation will contain conceptual representations, signalling the extreme unpleasantness that gets associated with a *negative* value in the affective store (see Chapter 9 for further discussion of values and taboos). This has the effect of signalling the harmful character of the schema. It triggers an avoidance response similar to what causes the avoidance of locations in the outside environment associated with danger.

## Fading Memories

### *We are Constantly Learning...and Forgetting*

We have established so far that activation has the effect of increasing the accessibility of representations. Forgetting is one part of the whole activation story. Inactive representations automatically experience a very gradual decline until reactivated frequently enough to arrest the decline and raise their resting levels. It seems to be the one cognitive process that can work with representations while they are *inactive* ('offline').

The positive aspect of forgetting is that, in all probability, it promotes processing efficiency. Lack of use, after all, means lack of usefulness, so that allowing resting levels of less useful knowledge to decline means that its representations cannot compete as well as they did when they were more useful. There will always be competition anyway, but fewer serious candidates competing at the same time and similarly high levels of activation will allow speedier solutions. You could describe this as 'clutter reduction'.

### *Lost or Buried?*

It still remains an open question as to whether we ever 'lose' something completely. In principle, the resting level of activation possessed by a given representation could fall so low due to disuse that, from the individual's point of view, it effectively becomes completely 'inaccessible', that is, in any schemas in which it happens to have been taking part. In other words, 'forgotten' would therefore not mean literally 'extinguished' but just that some knowledge has become practically impossible to activate sufficiently, at least for the individual to become *aware* of it. Once activated, the relevant representation(s) could still be affecting processing. There is also the possibility that, at some point and in some contexts, they could be reactivated up to levels that allow the individual to become aware of the knowledge that once seemed lost for good. Age regression hypnosis has opened up this possibility by allowing people to access long-lost memories from their childhood. However, since the mind can create false memories (adjusting original schemas until they represent things that never happened), the evidence emerging from such sessions needs to be critically scrutinised.

### Forgetting as Development

As time passes, certain things become less frequently used and they can then appear to us as being less easily accessible. Situations can then arise when we begin again to activate the affected representations(s) more frequently. Whether or not some deliberate intention or desire to recall something to mind is involved, it is actually the direct consequence of more frequent activation that automatically leads to the establishment of higher resting levels. The raised value of something that seems forgotten will give it an extra boost to its current activation and a subsequent rise in its resting level, as will be explained in more detail in Chapter 9.

As resting levels decline and rise, forgetting could be seen as just another type of continuing development, one that can go either way. This idea is based on the idea of ongoing change in the history of a representation and its networks of association. Memories and the schemas that support them one way or the other change over time, sometimes slowly, sometimes quickly, sometimes becoming stronger and sometimes seeming to fade away.

### False Memories

Those investigating a crime and calling on witnesses to report what they saw will be very familiar with the way in which people's memories change. What people *think* they saw and could perhaps have reported more reliably a few seconds later, the schema that was activated at the time of the event (perhaps even already flawed), will not necessarily be fully reproducible in all its details half an hour later. What may *not* change is their confidence in its reliability: 'yes, I am sure he was wearing a *green* t-shirt'. In the same way, an event in the distant past that made a great impression on you may be recalled 'faithfully' by you years later, but still has changed more radically than you realise. Some striking elements in the current schema may still match the memory you retained shortly after the original occurrence of the event. However, other elements would have fallen away. What were less salient aspects of the original schema will therefore have been 'replaced' with other stronger representations. Memory is nothing if not creative and driven by the need for coherence. This filling-in of gaps in a threadbare schema happens subconsciously. The more it happens, the greater the discrepancy between earlier and later versions of the schema used to recall the event. However, your unfounded confidence in the reliability of your memory may remain, and so the positive value and conceptual label (true) you place on what you recall may not change.

A dramatic example of the discrepancy between the belief you place in a memory and its relationships with some past event is a *completely* 'false' memory. In such cases, it looks as though the mind has really fabricated the whole event being recalled. Pierre Janet and then Sigmund Freud studied this phenomenon, and later still, the American psychologist Elizabeth Loftus carried out a lot of research into false memories. One of them, *Eyewitness Testimony*, provides examples of the many witnesses who differ in their reports of a crime while expressing equally high levels of confidence.[7] Then there is the 'Mandela effect', which relates to a false memory that gets spread around and is shared by many people. In this case, it originated in a radio programme hosted by Art Bell in the early 2000s, with callers all claiming they absolutely remembered Nelson Mandela dying in prison in the 1980s. In fact, he died a free man in 2013.

## Forgetfulness in Later Life

In line with what was mentioned in Chapter 6 about forgetting, as representations become more accessible by virtue of being frequently activated, knowledge that remains *unused* for long periods can gradually undergo a *decline* in its resting levels. Representations making up a particular memory schema will become gradually less competitive and harder to 'access' even when on a given occasion you might try to activate them: 'Now where was I at the time?', 'Did I ever visit that place when I was in the country?', 'Wasn't there some reason I never saw her?' and so on. In this way, you may forget words, names, visual memories of places you once visited, directions to locations that have not been required for a long time and so forth. Part of a currently available and activated schema may be intact while the resting levels of other associated representations may have declined enough to resist activation to the levels required: you remember a lot of the context but not the details you are interested in recalling. Without anything to keep them alive, memories can appear to weaken and even seem to fade away completely.

### Ageing Effect on Memory

Faded memories may often resurface, especially in old age. This memory revival effect is especially true of still remembered *pleasant* memories, but it also happens with memories of *unpleasant* situations, which the passage of time has rendered more distant and so less negative and less harmful. Unpleasant memories can still be awakened by virtue of, say, some disturbing event or person in the past being associated with them. Even taboos placed on suppressed memories can presumably get weakened because of disuse, rendering them less effective.

One striking example of the persistence of representations in their stores comes from the linguistic literature. Long-time Dutch immigrants in Australia in their later years were studied by the linguist, Michael Clyne, in the 1970s. They were, to their great surprise, experiencing their native Dutch coming back to life when they had fully accepted it had long been completely lost.[8] The apparent restoration in old age of resting levels of activation to competitive levels[9] challenges the 'use it or lose it' principle if, that is, by 'lose it' is meant 'extinguish completely'. The contents of the original schemas, which had become somewhat threadbare and fragmented, now begin to undergo restoration as more past situations become accessible again.

Not surprisingly, older people complain much more about a general loss of memory concerning things that are still of use in the present. Frustrating memory failures in old age may be related to the individual's experience of the complexities of modern life compared to what they were us to. However, this frustration is to some degree compensated for by this reminiscence effect, with the individual's faded past history becoming fresh again with newly remembered events that are pleasant to recall. They also took place at a time when life was easier to cope with for that individual. There is some evidence that this can be exploited to stimulate the cognitive abilities in the aged in the present. For example, an experiment whereby the subjects were immersed for a time in an environment that replicated their lives as lived many years earlier, in the 1950s, resulted afterwards in marked improvements in their mental and physical condition.[10]

### A Threadbare Adam Smith Schema

Mention was just made of memory schemas being 'threadbare'. By this, I meant that some parts of a fading memory schema may remain intact while other parts experience a decline in their resting levels. This is because representations that participate in such a schema each have their own activation history. Some will get more frequent use because they participate in other schemas that are more regularly activated as well. When you make a special effort to remember a name like Adam Smith, for example, you may notice that some bits of that memory do come to mind, but not enough to call up all the other members of the memory schema in question. You may remember a location, a person or an event associated with Adam Smith, or a sound or the first letter ('A') that the word begins with or even the whole of the first word ('Adam'). Some brief, deliberate recycling of these items in your mind may eventually trigger enough activation in the associated but still elusive missing items to make them grudgingly rise to the surface. 'Ah, I've got it at last. It was *Adam Smith*', you exclaim triumphantly. Or instead, that recycling process might continue subconsciously when you have given up all attempts to remember the full name, and your thoughts have moved on to other matters. 'Hah! Adam Smith', you suddenly exclaim out of the blue several hours later, and whoever you happen to be with will look at you in surprise, not understanding the relevance of what you have just shouted.

Finally, there is always the issue of changes in the *hardware* affecting recall. For example, apart from the fact that older people have so many memories due to their extensive life experience, resulting in more competition when given representations and schemas are activated, old age may bring hardware-related changes. This may result in, for example, faster rates of decline in resting levels when representations remain unused and the need to increase frequency of use where needed in order to compensate for this faster 'fading' effect.

This chapter has applied the general framework used to describe the mind as a whole to review various common ideas about memory, remembering and forgetting. It is important to keep in mind that these are phenomena that, however they may be described using everyday language, reflect the same general principles that govern the activation and storage of knowledge.

## Notes

1 See for example Storm et al. (2023).
2 Handley et al. (2015).
3 Kahneman (2011).
4 There is another intriguing example related to decision-making where a conflict arises between consciously invoked knowledge and related knowledge that plays a subconscious and complicating role but this will be explained in a later chapter.
5 Freud (1997).
6 Dodier et al. (2013). Recovered memories of trauma as a special (or not so special) form of involuntary autobiographical memory.
7 Loftus (1979).
8 De Bot and Clyne (1994).
9 It is not clear how extensive this return of the native language was, but even modest degrees of reemergence are significant.
10 Langer (2009).

# 8 Making Meanings

## Introduction

Chapter 8 discusses meaning representations and how they are constructed according to conceptual principles, managed online and stored. Meanings (conceptual representations) are associated with representations in a great many other stores. Once a meaning is activated in the conceptual store on the inner ring, activation instantly spreads out to all the representations that happen to be associated with it. Many schemas include conceptual representations, and, typically, multiple associations in a schema will end up intersecting at the conceptual store. This store, therefore, often functions as a crossroads or central 'hub' with spokes radiating out to other stores. Schemas, including conceptual representations, are regularly involved in processing at subconscious levels of activity. The same representations can also participate in a conscious activity when their status in the current context warrants it. The conceptual system is separate from but richly interconnected with language: not all meanings formed in the conceptual system can be efficiently expressed in words; examples include meanings associated with complex aesthetic experiences derived from, say, music or art. In any case, conceptual principles are different from those that specifically handle grammatical structure. However, they are still directly involved in shaping the communication and interpretation of messages, whether or not language is being used at the time. Various examples of schemas with conceptual representations are provided.

DOI: 10.4324/9781003606536-8

## The Meaning Crossroads

In this chapter, we take up our tour of the rings once again. Here, and also in Chapters 9 and 10, the systems under discussion are on the *inner* ring. The conceptual system is an extremely important system and one that has already been mentioned on a number of occasions in earlier chapters and will certainly figure in chapters to follow. This chapter, however, focuses exclusively on this meaning-making system itself (marked c on the inner ring). This system is one that has much to do with thinking and the many products of thinking and emotions. A much narrower conception of what 'mind' is would treat knowledge as the product of the meaning system alone. This would be mainly because most of us seem to associate mind with what is most familiar to us: thinking, which is conscious and perhaps even just logical thinking, so quite separate from affect, that is, our emotional lives.

As will be shown later, the conceptual system is also intimately bound up with the phenomenon of human language. It is the system that, given its amazing development in our species, might arguably be said to distinguish us most clearly from our nearest relatives with which we otherwise share so very much in common.[1] As already mentioned, meaning representations are frequently involved at the crossroads where all the other associations in a schema meet up: this makes the store of the conceptual store act like the hub of a bicycle wheel with spokes radiating out to stores on either of the two rings. The 'hub and spoke' image is also used in the recent neuroscientific literature, indicating, despite some longer-established views to the contrary, that there is a functionally independent meaning system in the brain as well.[2] There is also an ongoing debate associating concepts with specific neurons.[3]

The conceptual system in this book operates like any other system. It follows *general* processing principles but also has its own set of *specialised* construction and online management principles: its representations are coded accordingly and located in the conceptual store. It is not an all-purpose cognitive system that can, in that sense, be labelled 'domain-general', covering all types of knowledge. In the present perspective, it is another member of the interactive family of 'expert systems'. That said, we can still point out something special about the conceptual system within this family, namely the way it connects up with other systems as a 'hub'.

## The Clear Difference

Fellow primates, for example, but also other species like octopuses and dolphins, may have their own equivalent of a conceptual system and even some ability to engage in abstract thought. Ultimately, most differences between species are always going to be nuanced depending on what characteristics are being considered (see Chapter 11). The conceptual system as it exists in modern humans is, however, one cognitive system that does separate us out very clearly from our fellow primates in two important respects. It does this in terms of the sheer *quantity* of conceptual representations that it can acquire, and secondly in terms of the *complexity* of that knowledge. The related question of consciousness will crop up here, but this topic will be looked at in much more detail in Chapters 12 and 13.

The conceptual system is certainly the resource that enables complex rational thinking. It includes the enormous number of conceptually based schemas that we have accumulated as a result of education and or picked up from life experience in general. Its contents include, amongst other things, the core of everything that we, our conscious selves, think of as 'what we know', and you might say that its ultimate importance lies in what it has enabled the human species to accomplish on this planet. Studying the development of this system from birth and especially its final stages[4] is one way of gaining insight into the sophistication of human conceptual ability.

Being aware of our conceptual superiority – if 'superiority' is really the most appropriate term – should not make us lose sight of what we share in common with fellow planet-dwellers. In addition, there is the question of what humans actually *do* with their advanced conceptual ability, not all of it praiseworthy. Creatures like the sociable and peaceable bonobo live lives that should make us all envious and aware of our limitations.

## Amodal Systems

If the conceptual system is not quite the domain-general system in the terms described above, it is an 'amodal' expert system. In other words, it is, like all other inner ring systems, distinct from and independent of any of the five perceptual systems. In humans

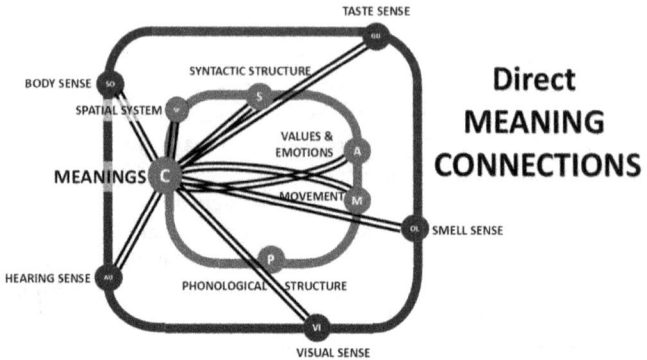

*Figure 8.1* The conceptual hub.

at least, meanings are not concepts that simply emerge from some combinations of representations on the outer ring. Take the meaning of a particular sound, like a dog barking, for example. This comes from an association between, on the one hand, an *auditory* representation on the outer ring (the *sound*) and, on the other, a *conceptual* representation (the meaning) formed on the inner ring.

The conceptual store is richly interconnected with most of the other systems in the mind on both inner and outer rings (see Figure 8.1). Consequently, activation in the conceptual system will immediately spread across the mind as a whole, involving representations in many other stores, which will very probably always include the perceptual stores on the outer ring. Even abstract meanings like 'truth', 'beauty' and 'independence' are conceptual representations that will typically have formed associations with representations of other stores that reflect examples of such concepts. Beauty, for instance, will be directly associated with the objects of visual and auditory experience that count as examples of beauty, such as given people, works of art, animals, sunsets and favourite pieces of music.

### Communicating Meaning

When communicating with others, for example, by speaking or listening, we are trying both to *express* meanings using any resources at our disposal and also to *interpret* what someone else is trying to communicate to us. The same goes for writing and reading or producing and interpreting sign languages. It also goes for *non*-verbal means of communication like facial expressions, body posture, icons, traffic signs, cartoons and other images.[5]

Although closely associated and involved with the meaning-carrying system that we call 'language', the conceptual system itself cannot be classed as a language or 'linguistic' system since, apart from its separate processing principles, it deals with *all* kinds of meanings including meanings that may not necessarily be easily and economically expressed in any language known to a given individual as well as the meaning of non-verbal signs like emojis and traffic signs, these being at least translatable into any language. Learning another language can also reveal words with meanings for which there is no word in the language(s) you already know, although they are understandable. Some of them might be intriguing or even useful to you like the French expression for thinking of a witty way of responding to something someone said when that person has already moved on to other topics or gone away: 'l'esprit de l'escalier' ('the wit of the staircase'), or perhaps the Japanese word for acquiring books for reading later or even letting them pile up without actually reading them, 'tsundoku'. Not so many people living further from the Arctic will find Finnish 'poronkusema' too useful since it is just a word referring to the distance a reindeer can travel without wanting to urinate. Other examples include feelings that lovers of music or art have when responding to a particular painting or a piece of music; they may still struggle to communicate their aesthetic responses in words. Connoisseurs of art, music and food will develop numerous terms for their own use to express subtle differences in taste, for example, but if you are not a connoisseur lacking this terminology, you may still be able to meaningfully discern and recognise many of these taste differences. It is difficult, nonetheless, to think of conceptual representations without thinking at the same time of *language* meaning. After all, we want to be able to communicate as many meanings as possible to other people, and language is really the best way of doing this. At the same time, not having a word for a particular meaning is no obstacle to being able to understand the concept it expresses.

It is still possible to communicate quite a lot without the use of language. This is something we can experience when communicating with people when there is no shared language available. We use sounds and gestures, and for example, point to get the meaning across. The rich relationships between meanings and language(s) in the mind should not deter us from thinking of the conceptual system as separate from the systems that determine the linguistic structure. Conceptual principles of construction are different from those that determine the possible ways in which grammars, including the grammars of sign languages, can be developed in the mind. It is still possible for us (and even chimpanzees) to use the conceptual system to put signs and gestures into a particular order, a kind of primitive quasi-grammar to convey a meaning, but these simple sequences are short and nothing like the elaborate ways in which we are able to construct long stretches of written, spoken or signed text to convey complex messages. Some people claim they do not think in words and sentences, but the meanings behind them can somehow be smoothly combined to express their thoughts, perhaps accompanied by images or a hybrid flow of words and images.

### You Say 'Semantic'. I Say 'Conceptual'

The conceptual system marked 'c' on the left of the inner ring in the metro map, first displayed in Figure 1.1 in Chapter 1, is also repeated with some added interfaces in Figure 8.1. Some people might want to label it with the more common term, 'semantic'. Is this the same as 'conceptual'? It could be. However, the term 'semantic' has a number of different interpretations both in psychology and in linguistics. This can be confusing. In linguistics, *semantic* meaning, the basic meaning of single words, phrases and sentences taken out of context, is often contrasted with *pragmatic* meaning, the meaning of words and also longer stretches of language used in a given context where their basic meaning can change. 'Red' and 'blue', for example, as in 'blue sky' and 'red hair' retain their basic colour meanings but in other contexts, they can mean something quite different as in 'he saw red' (became uncontrollably angry), 'the blues' (a type of sad music with a strong beat) as well as red and blue used to denote contrasting political views. 'Is anybody sitting here?' is a seemingly nonsensical question but directed to someone about an empty chair next to them in a restaurant or train has a pragmatic meaning equivalent to, 'Is this seat taken?' In general psychology and, especially in everyday usage, semantics can be a more inclusive term simply indicating meaning in general. Given the ambiguity of the term 'semantics', 'conceptual' is the better and most inclusive way of describing a cognitive system that deals with processing abstract meaning.[6]

### *Making Many Things Meaningful*

Unconnected meanings kept isolated in the conceptual store would really be of no use to man or beast. The associations with the contents of *other* stores are crucial. In other words, it is what meanings are connected to that really counts: a meaning must be a meaning of *something*. Abstract meanings could be an exception to this rule and, in principle, need no associations with anything outside the conceptual store itself. As suggested earlier, they will, in practice, get 'anchored' in the mind via associations that *do* involve representations in other stores. This also suggests that without a connection to the conceptual system, everything else we know one way or another, whatever other kind of knowledge we are talking about, is quite literally 'meaning-less'. There is, of course, nothing strange about this idea: everything outside the conceptual store is meaningless unless a meaning has been ascribed to it. The 'orange' example can again be used to illustrate: the visual representation of an orange *by itself* has no meaning. It may produce a visual experience of an orange in your mind, but that alone does not tell you anything about it. It needs to have a meaning association to become 'meaningful'. Very often it will have associations with sounds, signs or text patterns (writing), so that the meaning can not only serve to shape thoughts but also be communicated to others. These visible and audible patterns will inevitably involve further direct associations with a particular type of *linguistic* structure, that is, an adjective or a noun in the syntactic system (the station marked 's'). You may imagine that there will also be *indirect* connections as they spread out in schemas, creating ever more complex networks of association. Figure 8.1 just concentrates on direct associations: those that run through pathways that connect the c store directly with stores of other systems. As you can see, there are already quite a few of them.

### Sensory Perceptual Meanings

Following on from the orange example, below, in Figure 8.2 are three more simple examples of what the conceptual system can do when meaning associations are formed, in this case between the conceptual system (c) in the inner ring and those outer perceptual systems lying around the outer ring. In the following three examples, each of the different perceptual representations (smell, hearing, taste) located on the left side of the interface connecting two stores, marked with a double arrow ($\leftrightarrow$) and written in uppercase letters, is associated with a particular conceptual (meaning) representation on the right side of the arrow. For the reader's convenience, the resulting meaning in each example is very loosely expressed in words underneath each pair. In all these informal descriptions, note that, for example, '*smell*' refers to the SMELL *concept* and by the same token, '*sound*' is the SOUND concept and '*taste*' is the TASTE concept and so on:

### Affective Associations with Conceptual Representations

As can be seen in Figure 8.1, there are two curved inner-ring connections. The topmost one connects the conceptual store with the 'value and emotion' store of the *affective* system (A). As mentioned earlier, both values as well as basic emotions are handled within this affective system. As will be explained in Chapter 9, each basic emotion is a particular combination of representations, and each combination always includes a component representation that is either a negative or a positive *value*. A given meaning (conceptual representation) in the conceptual store

**OLFACTORY $\leftrightarrow$ Conceptual**
(roughly) "*this is an orange smell*"

**AUDITORY $\leftrightarrow$ Conceptual**
(roughly) "*this is the sound of my doorbell*"

**GUSTATORY $\leftrightarrow$ Conceptual**
(roughly) "*this is a salty taste*"

Figure 8.2  Sensory meaning associations.

could, for instance, become associated with a *negative* value in the affective store. It could also become associated with an emotion like, *disgust*, for example, in other words, with an affective representation that would anyway include, in this case, a negative value. Figure 8.3 illustrates some conceptual/affective associations. A single meaning/value association is followed by a triple smell/meaning/value set of associations. Again, '*smell*' is not the olfactory sensation but the SMELL *concept*:

### The Meaning System in Humans

So far, at least, the (relatively) neat and ordered way in which cognitive systems in the mind have been presented might still stand in sharp contrast with the complexity of the ways in which the human *brain* operates. Describing the details of what operations in the *mind* are made possible, despite the simplicity of its basic architecture, may already be revealing an impressive level of complexity in its own right. The more we look at the conceptual system, for instance, the more amazing, and indeed complex, that system turns out to be. Over the lifetime, humans will accumulate

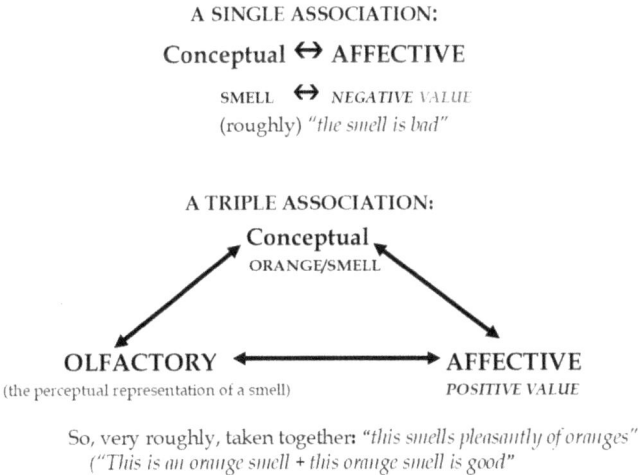

A SINGLE ASSOCIATION:

Conceptual ↔ AFFECTIVE

SMELL ↔ *NEGATIVE VALUE*
(roughly) *"the smell is bad"*

A TRIPLE ASSOCIATION:

**Conceptual**
ORANGE/SMELL

**OLFACTORY** ←——————→ **AFFECTIVE**
(the perceptual representation of a smell)     *POSITIVE VALUE*

So, very roughly, taken together: *"this smells pleasantly of oranges"*
(*"This is an orange smell + this orange smell is good"*

Figure 8.3 Conceptual associations: three examples.

a truly amazing number of meanings, many of them extremely complex because the knowledge repertoire in their conceptual stores is continually expanding.

It is possible now to claim – fairly confidently – that humans do have an independent conceptual system in the sense described in this book. It is less clear to what extent the same applies to other species. Possibly primates, for example, have some sort of independent conceptual system, but for many other species, it just seems less likely that they would have one. When non-human parents observe their young with obvious affection or when certain objects provoke fear, we can talk of schemas linking sensory perceptual representations with representations in what would be their own inner ring systems, notably the affective system. Although there may be no conceptual system as such available, it is still legitimate to use 'meanings' and 'meaningful' in a more informal sense to describe certain aspects of animal behaviour.

We are not done with the conceptual system yet and will certainly be returning to it in later chapters. Right now, since it has been mentioned several times in this chapter, it seems an opportune moment to look more closely at value and emotion; in other words, at the all-important and often underestimated human affective system.

## Notes

1 This claim will be reassessed when we come to the Chapter 14.
2 For those interested in this approach, see Ralph et al. (2016).
3 Quiroga (2012).
4 'Stages' here are in a general sense so not reflecting the pioneering work of Jean Piaget, whose stage theory work is rich in insights but also, for example, dependent on studies generally considered insufficiently rigorous. (Oakely, 2004).
5 Cohn (2013).
6 Jackendoff (1990).

# 9 Value and Emotion

## Introduction

We now turn to the affective system on the inner ring (marked A on the map). This system deals with what are generally known as 'basic' emotions, although their exact description differs slightly depending on the adopted approach. Basic emotions appear not to be the sole possession of humans; their outward manifestations, easily recognisable in many other species. This system also includes positive and negative values. That can be associated with many different representations, and its store has many pathway connections with other stores. In our imagination and in fiction, characters often struggle to decide between two classic options, one the 'sensible', 'logical' choice and the other guided by their emotions. In reality, given the way that the patterns of association in so many schemas involve both conceptual and affective hubs, the common distinction between 'heart' and 'mind' – or 'emotion' and 'reason' – has become impossible to maintain.

## Simple and Complex Affective Representations

One way or another, depending on the theory, affective representations include a set of basic emotions like fear, joy and surprise, and, however they may be defined, can be counted as part of the affective toolkit. More subtle emotions are associated in particular with humans, although some other species might be included. Emotions like nostalgia – a longing for a fondly remembered past – guilt and gratitude, for instance, will involve associations with representations in the conceptual store and at least one basic emotion.

DOI: 10.4324/9781003606536-9

As an affective representation, every basic emotion, like fear, includes in its makeup either a positive or a negative value: these are also affective primitives. With fear or anger, it will be a negative value. Each basic emotion is already a combination of two representations since it will also have an associated value. Values are also contextual influencers in that they can also be independently associated with many other types of representation and can increase default activation levels in response to a given context. This means that in some situations, particular representations can become more highly valued than they would be otherwise. Their association with a value in a given context will give a boost to the representation's activation level, and this boost will spread across to other representations in their schema. In other words, when something changes in the situation, the current value associations of given representations in a schema may change, with different representations getting boosted in response to the changed situation. The mind's adaptation to new situations will often have the effect of changing its current priorities, and this adaptation will be reflected in a reassignment of value associations to reflect those new priorities. Value associations are therefore not only those provided in advance or those that have already been formed by experience, but also new associations that are formed on the spot in response to shifts in the current situation. In all cases, the value, be it negative or positive, can undergo shifts in its current activation level in an ongoing, changing situation, making the value more or less influential. In other words, value associations do not have to disappear completely in a given situation, but they can fluctuate.

## The Affective Hub

Along with the conceptual (c) store, the affective store has multiple direct connections with other stores. As such, it also qualifies as a major crossroads point for a large number of schemas with their various associations intersecting there. For this reason, we can speak of both systems, A and C, on the inner ring, acting as major hubs during the activation (formation and management) of a great many schemas. Relative to other systems, the repertoire of representations in the affective store does not change much. What happens over the lifetime is probably more a matter of creating and modifying associations between affective representations and other types of representations, although ongoing experience might lead to new combinations of basic emotions within the affective store.

### Affect, Emotions and Feelings

It is difficult to overestimate the importance of the affective system and how it drives human behaviour and thought processes. Emotions are commonly described as different kinds of feelings we have, feelings that we are consciously aware of and can describe in any language. To do this, we use a rich repertoire of words like English 'down', 'low', 'gloomy', 'sad', 'contented', 'lethargic', lazy, 'elated', 'enthusiastic', 'energised', 'worried', 'nervous' or 'frightened'. To be a bit more precise about the terminology, 'emotion' is a more general term than 'feeling', although in everyday usage, the two words are often used interchangeably. 'Feelings', here, describe only the *experience* of emotions like the feeling of fear or of sadness. In its turn, 'emotion' itself forms part of 'affect', the latter term applying both to the system and to representations in its store.

### Emotions and the Subconscious

Emotional states can sometimes go largely unnoticed in our subconscious, or otherwise, we can be aware of an emotional state itself but not be able to fully identify it and understand its cause. In the second case, we are then conscious of the feeling, the physical sensations associated with it, and whether it is negative or positive, but can associate no specific meanings (conceptual representations) with it. We cannot identify exactly which emotion we are experiencing, that is, which feeling, and we cannot understand why we feel so good or so bad. People can be sad about something or depressed or jealous without fully realising it. They can suddenly find themselves attracted to, or suspicious of, something for 'no good reason'. Being aware of having some feeling, especially one that appears to have no obvious cause, can sometimes be disturbing. There may be a persistent need to understand or control it. In other words, they try to rationalise it. 'Me depressed? No, of course not. I've probably got some sort of virus that's running around'.

The discrepancies between what we think and the hidden drives that motivate our behaviour are a rich field for politicians and advertisers to exploit. They are well aware that they can manipulate people subtly to adopt views and preferences to their political or commercial advantage. As a result, people may become attracted to particular political parties, people or products on the market that they might otherwise have ignored. We can also be manipulated in the opposite direction, making us suspicious and likely to avoid things that formerly did not bother us in any way. In this sense, the disinformation that is spread around the internet is a pretty crude device compared to other long-established types of trickery. In any case, the question is what is going on in those subconscious depths of the mind to influence people against their better judgment?

### The Heart Versus the Mind?

Scientific studies of emotion, once viewed as quite separate from the bases of rational cognition, are now treated as part of a single story. This is true even in brain research, where *affective neuroscience* (a term coined by Jaak Panksepp[1]) is regarded as part of the general study of cognition. The underlying message now is that the rational and emotional, the 'heart' and the 'mind', are too intertwined to be able to treat them as polar opposites. To the extent that we can still regard the conceptual system and the home of rational thought and the affective system as the place where the 'heart' is housed, it is impossible to see them as operating in isolation from one another. They may be in different hubs, but in reality, schemas will typically contain representations from both systems: the webs of association that radiate from each of the two hubs get truly intertwined.

### Ideas

At the heart of each of these various emotions lies what is sometimes part of what is called 'appraisal'. Appraising and feeling might also seem to be distinct processes. Appraisal refers to the ways ideas, states, people, objects or events can be interpreted and associated with a particular value representation. We naturally think first of conscious assessments, but appraisal does not have to be a conscious process and more often isn't. Values can be associated with practically anything stored in other systems. This is what was meant earlier by values being regarded as 'influencers' in the mind's response to whatever the current context is, also when that context changes.

## Affective Biases

Appraisal is something that is going on all the time in the subconscious parts of our minds. Often, as just suggested, there is a conflict between the two ways of judging something. There may be something that, after some conscious reflection, we have evaluated as positive and desirable, but, deep down in the recesses of our subconscious minds, it may still be valued as negative and to be avoided. Inner conflicts between subconscious appraisal and conscious evaluation may make our behaviour seem to us inconsistent, irrational or worryingly unpredictable. Someone can feel antagonism towards a person with a different ethnicity, but at the same time, have the firmly held belief that such antagonism is unjustified and wrong. It would be easy to think of those holding racist views in a simplistic way ('either you are a racist or you have not a racist bone in your body'), according to how these feelings are expressed. The trouble is that we all have an inbuilt tribal instinct that has long been valuable for survival, one that makes us suspicious of unfamiliar groups of people, 'outsiders'. Whether we like it or not, we may be biased in one way or the other, including in ways we disapprove of. Deliberate counter-measures can be adopted to inhibit a bias that is influencing our feelings and or outward behaviour. But even then, it might take some time for our affective associations to adapt in accordance with our intention to become truly unbiased not only in word and deed but also in our thoughts, if that is possible. To achieve that, there has to be a desire to develop a better understanding of the people we distrust, making them seem increasingly less alien. Sometimes, it might require a deliberate intervention to encourage changes in how 'outsiders' are perceived. Then, in principle, the bias and distrust will weaken or disappear completely. This is where some aspects of our DNA and the mind's biological starter sets can turn out to be eitheronly initial or otherwise persistent obstacles to the formation of larger, more inclusive groups and thereby to wider social stability.

Although the brain's emotional systems are often claimed to extend more widely,[2] the neural instantiation of basic emotional responses has long been associated with the limbic system, a set of structures in an evolutionarily older part of the brain. It is therefore not surprising that we recognise similar emotional responses in other species by the way they react physically to given situations. This ancient brain system forms what you might call a basic survival resource. The same can be said of the mind's affective system. You may remember, in Chapter 5, the example of ready-formed associations, that is, the snake-like shape example, where cats without prior experience of real snakes will respond defensively, or even leap away from objects shaped in a certain way. The fear of snakes and spiders in so many humans provides further illustration of the same idea. Snakes and spiders themselves are not known in advance as such, but the responses to certain related shapes (and movements) are already in place and do not need to be learned, only, where desired, unlearned.

Two examples of an association between a representation in the affective system and one in another system were given in Chapter 8, and they are reproduced now in Figure 9.1. The first simply associates the concept of TASTE (which itself is associated with the appropriate taste representation on the outer ring – not shown here), with an affective representation, just a value and, in this case the value, NEGATIVE. This just informs the individual that this particular taste is bad and should be avoided. However, it is difficult to imagine a negative association not being accompanied by *any* emotional response. The second association accordingly links the TASTE representation with the basic emotion, *DISGUST* (which includes its negative value):

### CONCEPTUAL ↔ Affective

TASTE ↔ NEGATIVE
(roughly) *"This is a bad taste"*

### CONCEPTUAL ↔ Affective

TASTE ↔ DISGUST
(roughly) *"This is a disgusting taste*

Figure 9.1  Affective associations: Two examples.

In this way, certain tastes may have a pre-formed negative *starting* value, a defence mechanism in order to help the child avoid eating or drinking things that may be harmful. To reinforce the child's defences, avoidance action will be needed; so, the simple two-way association will not be enough; so, a wider network of associations will usually be needed. This schema will certainly involve associations with the motor system, causing the child to withdraw from the source of the taste: the resulting instinctive movements will often include the facial muscles around the nose and the mouth contracting into a classic expression of disgust. In such cases, only experience and a subsequent change in value can alter their instinctive preferences and accompanying reactions. The child's reactions can, of course, continue into adulthood unless later experience causes shifts in value and associated emotions. Spelling out instinctive responses in the brain, in terms of the mind's operations in this way, can show not only the corresponding psychological mechanisms involved but also serve as one more illustration of the principles that govern all mental processing.

### Addiction and Motivational

Affective associations with other systems come in all shapes and sizes. Particularly striking is the case of addictive behaviour, whereby even an initially unpleasant taste or smell (for instance) is accompanied by unexpected pleasurable sensations. A neural account of this development would include a mention of the brain's 'reward-system' involving the release of the neurotransmitter dopamine.

Addictive behaviour is a complex motivational phenomenon. Although it is associated with the notion of excess, it does not have to be harmful. Apart from the associations that are connected more directly to the rewarding physical sensations themselves, highly valued associations can also be formed between the circumstances that give rise to the rewarding experience. Such circumstances include pleasurable socialising with other addicts that accompanies the direct expression of the addiction, but there are many more. Imagine the rich schema of associations containing representations that are most closely associated with the addictive experience. Added to that are all the representations associated with such circumstances that get coactivated when the activation of any one of them is triggered. These are often referred to using terms like 'psychological' dependence or 'psychological' addiction', and they involve the affective system and values with very high resting levels. Although they have different defining characteristics that mark them out as behavioural complaints, unlike, say, heart and liver disorders or viral infections, they are all ultimately *physical* phenomena as well, ones with neural features, many of which may not yet be properly understood.

### 'Heart Versus Mind' Conflicts

Because of its role in associating values with practically anything else in the mind, the affective system represents a powerful force that motivates our behaviour in all kinds of ways. It is probably the most powerful force in the human mind. Its influence can often work for the common good and for the benefit of the individual, whereas a dysfunctional affective system in an individual may be associated with sociopathic and psychopathic behaviour. In general, there is often good cause to be wary of the power of affect. Our conscious reasoning ability, based in our conceptual system and which we would like to think of as our strongest ally for coping with life, often seems to lose out when there is a conflict between what we like to think of as 'reason', on the one hand, and, on the other hand, emotion. The trouble is that what we like to think of as logical, rational thinking is, as already mentioned, not an entirely independent ability, free of influence from the affective system. This is because even our so-called 'objective' reasoning is not immune from assumptions and beliefs that have already been given positive or negative values, and which the individual may not even appreciate. However, the actual logical process of coming to some decision may be like, for instance, choosing a candidate in an election, or a partner or a particular belief system; the completely rational basis of our decisions, their underlying assumptions, may be questionable. Already existing assumptions and beliefs will lead us to highlight any supporting evidence and ignore or suppress the awareness of disconfirming evidence when it is found to be a threat to mental stability and the quality of an individual's life. This is what cognitive behavioural therapy (CBT) tries to disrupt: therapists try in various ways to bring these underlying assumptions and beliefs into the open in order for them to be challenged and changed. This is also what makes it particularly challenging to conduct a purely rational debate about opposing ideologies, be they political, philosophical or religious. It is because, for the most part, individuals are not arguing on a level-playing field. It is difficult to disentangle what can be freely and openly debated from those (sometimes unstated) assumptions that are rooted in deeply entrenched emotional biases. Politics is full of such exchanges between people. Each idea presented in a debate comes as if it were with (often undeclared) affective baggage, such that the discussants may appear to be talking about the same thing in the same way, but, in fact, are not.

Meanings and affect can be bound together so tightly that it once led neuroscientist Joseph LeDoux to reflect that human beings at this present stage of evolution have yet to evolve a satisfactory, balanced relationship between their emotional system and their rational ability.[3] 'Heart' and' mind' are still in a state of imbalance, and this may often have negative consequences both for the individuals and all kinds of interactions between individuals singly and in groups. We get to see this heart–mind imbalance most starkly amongst people in their teens and early 20s. Parts of the brain associated with logical reasoning in the prefrontal cortex are not fully developed until the mid-20s, and the effect of emotion on behaviour may well be even greater than in the mature adult: partly dependent on individual temperament, higher degrees of irrational behaviour, mood swings and risk-taking can be relatively more prevalent amongst members of this particular age group. That said, emotion can, of course, equally motivate rational decisions in a positive sense as well.

Here now, in Figure 9.2, is the metro map again, but now just focusing on the various connections (marked as always as double lines) radiating directly from the affective system (marked A in the metro map) and linking it directly with six other systems. It is this network of (double-line) pathway connections that serves as the basis for multiple schemas to be formed and get activated. You may notice that the two systems responsible for language structure do *not* have a direct connection with affect, a topic that will be returned to later, in Chapter 15.

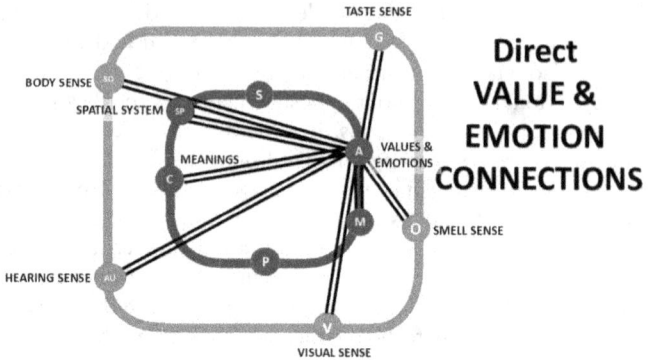

*Figure 9.2* The affective hub.

### Tigers and Cocktail Parties: Examples of Boosting

Affect in the mind (and brain) has a direct impact on our behaviour in a particular way. When an affective representation that involves a positive or negative value is activated, the level of anything that is associated with that affective representation will get a boost as well, although the boosting effect that spreads across other associated representations may be uneven. Imagine an electric current suddenly emanating from the affective representation to anything it happens to be associated with and imagine the resulting power surge across the whole schema in question. The sight (visual representation) of a tiger running in your direction will trigger a particular version of what we could call a 'tiger schema'. Experiences with tigers and the schemas they evoke could be varied, ranging from traumatic to, say, almost endearing when encountering, from a safe location, a beautiful tiger, its eyes closed and fast asleep. Affective boosting is determined by the appraisal of whatever happens to be the current (internally represented) context. In the first case – your representation of the tiger rapidly approaching you – a major surge will be transmitted to many representations, including some in your motor system, all of which will be part of a fight-or-flight response. Other affective features in this schema that may be activated along with the visual image and the concept of tiger might be associations such as the bright colours of the tiger's coat and the beautiful grace of its movements. These visual features and associated values, however, will not be relevant in the perceived circumstances and, therefore, will only be weakly activated if at all. Because of their survival value, the tiger-related associations that are strongest, that is, those that already have very high resting levels, will anyway be those that are most associated with danger, and which trigger the basic (negative) emotion, *fear*. Values are always activated as part of a whole schema of contextual associations: as always, the current internal context exerts a powerful influence on what gets activated and, in each case, how intense the activation will be. Furthermore, since values have this boosting effect, emotional states are always accompanied by some level of awareness. At the very least, the individual is aroused and aware of this state of arousal and also of its pleasant or unpleasant character, along with an awareness of what is directly causing it, if the source is easily identifiable at the time.

One other example of affective boosting is when you are in a busy room talking to someone with lots of noisy chatter going on in the background. This explains why the response in this type of situation has been dubbed the 'cocktail party effect'. Part of the total context that your mind is dealing with is currently in the background, that is, of less value, whereas another part has a raised value, being currently of greater value to you: you are concentrating on what someone is saying and ignoring the buzz of conversation around you. Despite the fact that you have seemingly blanked out this background chatter and are attending to what one person is saying, when your ears pick up a mention of your name by someone talking in the background, suddenly your attention is switched away from your interlocutor and towards the source of that mention. This switch was not the result of any deliberate (conscious) intention: it happened automatically. Why? This was because the sound of your name is very important to you. It is associated with a meaning (ME[4]), already possessing a very high resting level. When activated in your auditory system, the auditory representation ('sound') of your name will immediately be co-activated not only with its meaning association but to the whole network of associations, including the positive value representation in the affective store. The boosting effect of this coactivated (positive) value will instantly spread increased levels of activation across the whole schema. The end result is the sound (auditory representation) of your name in the background suddenly dominating the voice of your current conversational partner. In this way, affective processing can play a major role in explaining what most people would call your 'attention'. Switches of attention can be triggered when something that has a higher value and hence gets intensely activated outcompetes what is currently the dominant representation.

### Complex Emotions and Guilty Pets

The basic emotions, their definitions and their number, vary according to which perspective researchers in this area adopt. Broadly speaking, they are the emotions that we see reflected not only in our behaviour but also in other species, such as *anger, fear, disgust, sadness* and *joy*. They are the same basic emotions that are located in our human affective store, all of them being minimally complex since they already have a value association. More complex emotions like *nostalgia, jealousy, guilt* or *pride* are probably combinations of those basic (positive or negative emotions) along with particular associated representations in the conceptual store that reflect various subtleties of meaning. It is just possible with other species that complex emotions may also be experienced – but without the same level of understanding – with the coactivation of more than one basic emotion within the affective store, for example hybrid emotions such as anger combined with fear and each emotion varying in their respectivet levels of activation.[5]

Following what was said in earlier chapters, activating any of these 'emotion' representations, like fear or joy, will create sensations of which an individual will be aware. They will have a particular feeling. But their significance as far as an individual's mind is concerned is not established until meanings get to be associated with the feeling in question. This happens via the connecting pathway (interface) between the affective system (A), on the one hand, and the conceptual system (C), on the other (see Figure 9.1). In this way, other species, say a dog, may experience the same basic emotion as a human, for example, joy, and physically respond to this feeling in easily observable ways. However, only humans will understand this feeling as joy as opposed to some other emotion. This allows a human to recognise and reflect on the emotions they are feeling, as when they reflect: 'I haven't felt this happy for a long time' or 'why am I feeling so depressed right now? There doesn't seem to be an obvious reason'.

Summing up, emotions and values like meanings always get associated with representations in other systems. In the case of emotions, most obviously, this includes somatosensory and motor associations. However, as just mentioned, conceptual associations with emotions permit ever more subtle labels to be added and used to distinguish a whole variety of more complex emotions. Note that animal lovers like to freely ascribe complex emotions and what causes them to their pets, for example. 'It' or 'she' is feeling jealous', or 'He's so guilty because he knows he shouldn't have done that'. This is very easy to do, but we have no idea whether or not it correctly describes the animal's emotional states. The network of associations that have been triggered by what the animal has perceived could just have created the sensing of threat connected with the presence of another animal, in the case of jealousy or a feeling of wrongness, without any understanding of exactly why the threat is being experienced in the case of guilt.[6]

Various complex emotions may, in given languages, have equivalent names that serve to identify them. At the same time, it is still possible for people to experience a complex emotion for which their particular language (or indeed any language) has no label. Furthermore, whether or not we have a word for a particular emotion or combination of emotions that have been activated, we will only become *aware* of feeling them once the affected representations have attained a sufficiently high level of activation to trigger that awareness. The awareness will probably not come directly from the affective system itself but will be experienced via the various associated sensations that emotions can induce. This indirect, sensory source of awareness will also be applied with the awareness of space and motion, the subject of Chapter 10. Prior to that point, they will be active in the underground, subconscious part of the mind, but already influencing our behaviour in ways that we will not appreciate at that time.

### Drives as Schemas

Finally, the place of instinctive drives like hunger, thirst, sex and avoidance of harm should be mentioned. These clearly involve associations with affective primitives, notably positive and negative values. They also involve goals which do not rely on experience and so come ready-formed. The question is where to locate the goals themselves. In the case of deliberately constructed goals, the straightforward answer is that 'goal' is a *conceptual* primitive. In a particular situation, it (GOAL) is part of a goal *schema* which details the intended target and the means for getting there. In the human mind, the simplest explanation is the same for instinctive drives, except that very basic goal schemas for each drive are provided *in advance*. These schemas will be further elaborated to fit the current context.

## Notes

1  Panksepp (2014).
2  LeDoux (2002).
3  LeDoux (1996).
4  This 'self' concept will be discussed in Chapter 15.
5  Plutchik (2002) and Eckman (2007).
6  In a podcast, Zoe Strimpel examines why so many people have become passionately obsessed with dogs. 'We have moved', she writes, 'beyond affection, beyond dog-is-person's-best-friend love, into a passionate confusion whereby we now seem to think and feel that there is literally no difference between pets and people'. https://bbc.in/3cGVG83?fbclid=IwAR2wI22 U18WSOuqZDPldIICmDRxgbOqVOSh_-7CM56 Lczogt8XzfOJHZLKM.

# 10 Space and Movement

## Introduction

This chapter will now cover two more inner ring systems: first, the *spatial* system, marked sp on the map, and secondly, the *motor* system, marked m. The spatial system covers the way we understand and use three-dimensional space. The motor system is the system that enables the activation and creation of what might otherwise be called motor 'programs' (in the form of complex motor representations) that serve to articulate our body in response to experience. Both of these inner systems, being one remove from states and events occurring in the physical environment, receive input coming from stores on the outer, sensory perceptual ring as well as from inner ring stores. The spatial system, apart from representing the spatial aspects of the body's current relationships with the environment, is also used in combination with other inner ring systems to create imaginary spaces which describe abstract relationships of many different kinds, often expressed in visual terms.

## Space–time

We live our lives in the space–time continuum. This is, anyway, our reality as we perceive it. Space and time go on forever, again, as far as we humans experience them. Whatever other dimensions might in principle also exist, we have evolved to live with these two. Exploiting the power of the human conceptual system, theoretical physicists and writers of fiction can entertain challenging notions of parallel universes, teleportation and time travel, but for understanding the basics of how the human mind works, the

DOI: 10.4324/9781003606536-10

priority is the more familiar world. For this, we need a spatial system for dealing with space and a motor system for dealing with motion. Very often, movements have to take account of our current location in space so that the two systems involved, sp and m, will be richly interconnected.[1] These two systems will not be discussed in great detail: I will just stick to the basics and set out how they function as two members of the family of 11.

## Space

### *The Spatial World*

This section will be about moving through real but also through *abstract* space. 'Real' space refers to the environment in which our physical self, as we understand it, is located. This environment is created in our minds initially via our sensory perceptual systems on the outer ring, including the somatosensory system that gives us the sense of our body position and movement. The space is called 'real' because that is the way we experience it and assess it. Ultimately, it is just an assumption, whatever the practicalities involved. Whatever Dr Johnson said to Bishop Berkeley about a stone, we should always treat reality as an 'assumed but probable' reality. Anyway, (so-called) 'real' space can range from our immediate surroundings to what lies many lightyears away into the far reaches of our universe. Beyond that is infinite time or infinite space, which is difficult to conceive except in very abstract terms. In more practical terms, there is what might be called the metaphorical space – in our minds. This is where we use our conceptualisation and experience of space, such as it is, to create, organise and use many different relationships between abstract concepts to give them some familiar properties and make them easier to think and talk about. For example, describing someone as being 'in' or 'out of' *control* or '*under* a lot of stress' would be two examples of this: more examples are provided later but the model of the mind here reflected in the 'metro map' with its two rings and 11 systems linked in various ways by an invisible network of pathways could be regarded as another example.

Knowledge needed for a proper understanding of the physical space around us is distributed across different types (auditory, visual, etc.) of representation located on the outer ring. However, these need a higher level of organisation to create a fuller and more coordinated sense of the space we live in. This can only happen if perceptual representations are associated with representations created and stored on an inner ring system, namely the spatial (SP) system. This is a system that we certainly share in one form or the other with other species that have the same need to navigate our shared planet. Humans, however, have an extra resource for understanding space, namely a complex conceptual system.

## Spatial Concepts

Our spatial system is, in its interactions with other systems, sufficient to cope with basic, everyday tasks related to 3D space, to locate objects and represent their spatial relationships with one another and also to identify physical obstacles (like rocks). To get a more comprehensive sense of the three-dimensional space that we inhabit, our human minds can also form multiple associations between the perceptual systems on the outer ring, the spatial system and now the *conceptual* system on the inner ring. This will assist our minds in the building of goal[2] schemas that enable us to navigate in more complex ways. One example would be choosing between alternative routes to get to a target that is not yet present (a place in the city or even a location in outer space). We know that animals, rodents for instance, that may well not have a conceptual system as such but can also store and remember complex routes through a maze. Having a conceptual system means we can enrich the route information with conceptual associations. We can store the conceptually enhanced knowledge of how to get from one place to another in the form of mental route maps in our conceptual store with appropriate connections reaching out to representations in the spatial and motor stores. These navigational schemas (mental maps) will be detailed when the locations are familiar, or they may be sketchy and incomplete when the location is unspecified or unknown. Spatial relations between locations will be represented there. Where a given navigational goal is being formulated, such multiple 'route maps' will contribute to strategies aimed at creating safe and efficient ways of how to go about looking for something or someone. By so doing, we would have then acquired all the associations to make sense of, survive in and fully profit from the physical space around us.

As already suggested, although the somatosensory system provides some information about the body's location and positions, it seems from research into spatial cognition[3] that the ability to understand and formulate strategies to deal with 3D space, is something that does requires a specialised system for that purpose. This is why a dedicated system in its own right has been included in our interactive family of systems. Apart from its primary purpose of coping with the spatial world, the same system is useful in other ways as well. It is also employed, together with the conceptual system, to build ways of thinking about how 'past/present/future' time is organised, a concept that is crucial in all kinds of ways but especially for planning ahead. The importance of the spatial system for our survival and for the way we conceptualise the world around us cannot be underestimated.

### *Natural and Artificial Navigation Aids*

How the details of this system could be elaborated within the present mind-based framework is naturally dependent on contributions from researchers in spatial cognition, not purely focused on the hardware. The same goes for research on the motor system. However, taking a brief detour from mind matters, in parallel brain-related research, the *hippocampus* always gets an important mention. This is in relation not just to the creation of representations (referred to usually as 'memories') *in general*, but especially in relation to spatial processing and navigation. The example of London taxi drivers often crops up in this context. As a result of their memorising, over a period of about two years, a multitude of locations and routes in order to pass a stringent test called 'the Knowledge', the right posterior area of the hippocampus in particular gets considerably enlarged.[4] It seems that these aspects of the human spatial system are centred on areas in the *medial temporal cortex* (which include the hippocampus). In practice, using navigational maps naturally triggers a brain-wide network of activation involving other, connected systems. The outward-oriented representation of what lies in the space *outside* us and associated with the medial temporal cortex has to be coordinated with another type of map; in other words, the 'egocentric' map, which represents the relative positions of our own body parts. This inward-oriented map is associated with another area of the brain, namely the upper, rear part (the *posterior parietal* cortex).

Switching to a GPS, nowadays, to navigate around the place is a mixed blessing. Interestingly, it has been reported that drivers with a reasonable idea of where they should be going but using the car's GPS system can be misled because some of their spatial knowledge is deactivated when they start relying on their electronic guide, which can sometimes turn out to be unreliable. Then, drivers may even ignore very clear evidence in their surroundings that *should* have led them to suspect that the GPS was actually taking them in the wrong direction. The result can be that they end up losing their way. They wouldn't have done so if they had relied on their own local knowledge rather than their GPS: if they had at some point taken a wrong turning by mistake, they would soon have become alert to unfamiliar features of the scene in front of them and quickly taken steps to get back on track.[5]

### Strolling Through the Past

Some spatial memories seem to be highly valued and, for some particular reason, remain so, retaining their value over very long periods of time, apparently regardless of their current practical importance. Many people find they can visualise walking round a cherished location, a previous home, for example, in the distant past, remembering a host of details. Their memories are certainly retained not for any obvious practical reason, but rather due to their strong emotional associations. Additional support comes from features that happen to be associated not *only* with such memories but also extend to other currently relevant contexts as well, and which consequently get frequently activated. The schemas involved will therefore include representations with high resting levels due to the associated positive values as well as other representations that happen to be frequently activated in other contexts. A much-loved location in the distant past can thus remain available, such is the power of the affective system for maintaining high activation levels, which are not (apparently) useful for day-to-day life.

*Spatial Metaphors*

In 1980, two linguists, Lakoff and Johnson, proposed that spatial concepts are also used as metaphors for organising other kinds of abstract concepts and this has also attracted much interest outside the area of linguistics itself. The 'in or out of control' expressions were examples already mentioned above. These are from English: metaphors will not be identical in all languages and all dialects. With regard to (English) examples based on the notion of time, the past is often conceived of as 'behind us' ('It's all *behind* you now. You have to move on'), and the future is conceived of as 'ahead' of us. 'Up' tends to be positive whereas 'down' tends to be negative. 'Up, up and away', the Jimmy Webb song sung by the group, Fifth Dimension in 1967 and used in Superman films, suggests a sense of freedom. Situations that 'go downhill' suggest decline: 'north' and 'south' are also used in this context, with 'south' being equivalent to 'downhill'. The idea of *containment* also crops up frequently ('staying within our remit', 'thinking *outside* the box', a limited '*envelope*' for making a decision, a '*window*' of opportunity, '*ring-fencing*' assets or a financial budget to protect them from risk and misuse and ensure it is used for only a selected purpose). The idea of protecting or securing by having something above you is also expressed using the concepts 'cover' and 'umbrella' as in 'this will *cover* your costs, insurance *cover*, *umbrella* contracts and agreements'. Note that, although these ways of conceptualising different aspects of reality are expressed through language, this is not about the way in which the linguistic systems (p and s) themselves operate but rather how the conceptual system works and its relationship with the spatial system.

The double lines in Figure 10.1 show all the directly connecting pathways that radiate out of the 'sp' system.

## The Motor System

This system handles physical movements of the body, and its main base is the brain for voluntary movements in the frontal lobe, where the primary and secondary motor cortices are located, but more generally, other areas elsewhere are implicated, such as the cerebellum and basal ganglia. The motor (m) system on the inner ring is not to be confused with the body-sensing perceptual

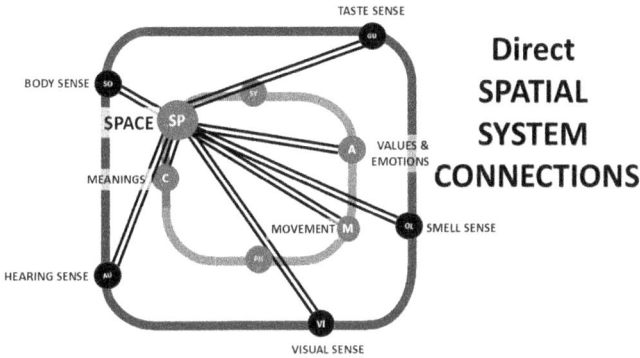

*Figure 10.1*  Direct connections with the spatial system.

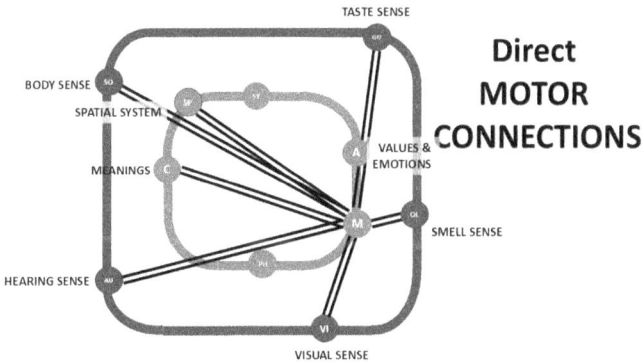

*Figure 10.2*  Direct connections with the motor system.

system on the outer ring, the one called the *sensorimotor* (so) *system*. Note, however, that the two stores of the systems in question are indeed connected via one of the five interface pathways that run between the motor system on the inner ring and the five outer ring systems (see Figure 10.2). We, as do other species, have a repertoire of *involuntary* movements like blinking, but also movements like those related to digestion and the circulation of blood. Less relevant for this account of the mind, perhaps, these movements are so important for survival that they resist full adaptation or suppression. What we will focus on here as most relevant to understanding the mind is the *voluntary* motor system in response to some current conscious or subconsciously conceived goal. For instance, motor representations can be extensively developed and modified when a particular skill is acquired. Some of these skills may require a precondition. The body must develop sufficiently first with walking, which develops naturally in the baby after about six months. The basis of this particular ability is not learned but inherited, even though developing the skill will also require some learned knowledge about the physical properties of particular objects in the environment that can support the process. By contrast, if the goal is to learn the motor skills for a particular sport or computer game, this requires the growth of representations that are responsible for programming new, coordinated movements. The movements they control require experience for their development: these are the result of the construction of complex representations, new combinations that were not provided at birth. The same is true of other species, of course, and they are a main topic in Chapter 11.

### Notes

1 McDougle and Hillman (2025).
2 Some details on goals are provided in Chapter 15.
3 Ishikawa (2021).
4 Maguire et al. (2000).
5 Javadi et al. (2017).

# 11 Comparing Species

## Introduction

This chapter discusses how humans compare with other animals. How different are we really from our fellow species? To what extent, when we imagine human-like thoughts and feelings in other species, are we just indulging in fantasies? Recent research certainly suggests that we have underestimated the intelligence of fellow primates and also certain bird and marine species. This indicates that we are not all that different from fellow inhabitants of our planet. Examining the differences and similarities can help distinguish those properties and characteristics that do make humans and the human mind uniquely different.

We can certainly detect the presence of awareness (consciousness) in other species. This has led some people to believe they can also 'think' at least to some limited extent. However, the most obvious sources of comparison must lie in the outer ring systems that help species navigate and respond directly to the physical environment. On the inner ring, the spatial, motor and affective systems will also have their equivalents in the biological software of many other species.

## Animals 'Versus' Humans?

Traditionally, humans have often been seen as fundamentally separate from the beings that we choose (in English and some other languages) to call 'animals'; this is despite the fact that 'animal' comes from the Latin word *anima* meaning 'breath'. Some religious beliefs explain this sharp distinction by saying that, unlike humans, animals may breathe but do not have either minds or

DOI: 10.4324/9781003606536-11

souls that transcend death. Some people also claim that animals, unlike humans, can neither be conscious (aware) nor be able to experience pain, at least as humans do. Some distinction is made between most vertebrates (who give clear signs that they do feel pain in some way) and most invertebrates (who do not give such recognisable signs). The science involved is not yet clear. On the other side of such arguments, there are plenty of people, animal lovers, who refute such claims or appear to do so by attributing to animals, and especially their pets, many human characteristics, which, frankly, science has yet to justify. Such justification is difficult to achieve when we will never know exactly how it feels to be a cow and a cat or a bat, but evidence can at least be collected to establish how probable or improbable these claims can be. Animals experiencing stress and pain seem easier to accept when recognisable symptoms are displayed. Certainly, looking at the research carried out by Frans de Waal, a great proponent of blurring the traditional human/animal distinction,[1] it is not only humans that have minds and experience pain, a point of view that is certainly supported from the perspective adopted in this book as well. The question is then more about other specific aspects that are *not* shared with other species.

### Anthropomorphism: A Belief or a Game?

Anthropomorphism, ascribing certain human characteristics to animals without any clear evidence that they actually have these characteristics, is an understandable temptation, especially for pet owners who would like to imagine their pets having greater understanding and more sophisticated thoughts and emotions than seems plausible. This leads them to over-interpret their pet's behaviour as confirmation of their ideas about them. Some end up believing what they are imagining, while others are simply indulging themselves in playful interpretations. The more human-like the pet's behaviour, the more we are likely to do this. It seems, then, that we have two contrary ideas floating around. Either all other animals are radically different from humans, or they are not. The question to be asked is, could the basic account of how the human mind works as presented in this book provide some guidelines for the scientific investigation of such questions and also the kind of arguments we have with our friends?

Recent research on primates and other species gifted with unexpected levels of intelligence, like birds in the *corvid* family (crows, ravens, magpies and the like), has certainly made many of us more aware of how much closer we are to our fellow beings than we originally thought. We can no longer see ourselves as so very special and distinct on this planet from all those fellow creatures around us. It is often more a matter of degree rather than a sharp distinction, and there are areas, as already mentioned, where we are very much *inferior* to other species, especially when it comes to the senses, physical strength and agility.

It is useful to begin with what systems, in broad terms, we have in common with other species, especially the ones nearer to us in evolutionary terms. We can then look at more radical ways in which we differ, especially in areas where we feel that *we* excel. We can do this without shying away from uncomfortable questions, such as 'why and how do we appear to be the most dangerous predator on the planet?' It turns out that, even given areas where we do seem to be markedly different from other species, we are still, from a biological point of view, 'fellow animals'.

## What Makes us Different?

Any systems in other species that have obvious human equivalents will nevertheless each have their own unique processing principles. Dog vision will still be distinguishable from cat vision, grasshopper vision, shark vision and, of course, human vision. When we turn to what most clearly separates us from other species and appears to establish us as the most cognitively advanced 'animals' around, two properties stand out, namely, a) higher levels of consciousness enabling thoughts (Chapters 12 and 13) and b) the properties that define human language (Chapters 14 and 15). Apart from various perceptual abilities, other species they may possess, there is no point in discussing the various ways in which humans are physically inferior to animals, for example, in terms of running, jumping, climbing and swimming abilities. These will be already obvious.

### Thinking and Sensing

Although this was not always the case, most people nowadays question the claim that only humans are conscious. This, of course, depends on what is included in the notion of consciousness. Although plainly aware of their surroundings and, in some cases, showing some ability to reflect and plan ahead, other species certainly appear to be much more rooted in the here-and-now and less able to engage in thinking and planning of any real complexity and in conceptualising hypothetical situations generally. Here, too, it is safer to talk in relative terms rather than adopt a black-and-white distinction: animals cannot think like us and are not conscious in all the ways in which *human*s are conscious. This leaves the discussion open to attributing forms of awareness to some species that may indeed include *some* degree of abstract thought if that is an acceptable way to characterise it.

Equivalents of the sensory perceptual systems being clearly present in many other species, each system builds and runs its particular kind of knowledge in distinctly different ways. The human olfactory (smell) system, as explained earlier, has its own unique operating principles compared with other human knowledge systems. Humans and dogs both have an olfactory system, but the dog's version, being infinitely superior to that of a human, will not be the same as ours. Again, some snakes have infrared vision. Eagles, clams, jumping spiders and snails also have very different visual abilities. The evolved hardware that all creatures are born with to suit their particular needs and environments inevitably means that their internally constructed worlds will be different, and their biological software will reflect this. In some respects, other species will, to our way of thinking, be superior and in others inferior to what humans possess. We should not, without question, take the particular version of the world which our own software has provided us with as more faithful in its internal representation of 'whatever lies outside us'.

### Activation Extended to Other Species

Finally, when it comes to activating knowledge representations, the general processing principles governing resting and activation levels should also work for other species that possess nervous systems. In fact, something like the metro map idea with its concentric rings could, in principle, certainly be adapted as a basic template to frame explanations and descriptions of other species as well.

# What Makes us Similar?

## *'Minding' Animals*

We are not used to employing the word 'mind' to indicate something that some species might possess. However, there is no reason in principle why we shouldn't use the word when talking about other animals, especially with those closer to us, provided we avoid unwarranted anthropomorphism. Readers will have to decide this for themselves once they have thought about all the various chapters in this book. For some readers, happy to attribute the notion of a mind to other species, it will be important to understand to what extent and in what way animals are aware of the world around them and what they make of that awareness. There is always the tricky question of definitions when debating such basic issues as mind and consciousness, and it is important to know exactly how the notion is being used. This can be no easy task in practice, given the wide range of opinions and ideologies that are supposed to help us understand the world we live in and that are brought to bear on such exchanges of view.

## *Awareness in Animals*

Certainly, both humans and other species share basic sensory awareness, but a common question is, 'Are animals conscious?' This question is very difficult to answer in any definitive way that will satisfy everybody: as Thomas Nagel[2] said, we have no idea what it is like to be a bat. It also depends on what you mean exactly by 'conscious' or 'aware', two concepts that are here treated as meaning the same thing (see Chapter 12). One thing is clear, though. Animals must be aware of what is going on around them, at least in some sense of the word 'aware': they are aware of their immediate physical surroundings, of their own internal states, and, in addition, they have the means of organising their behaviour in response to this awareness. This is as crucial for their survival as it is for ours. And, as it is with humans, at a given moment, certain representations of the animal's external world can get to dominate over others, allowing them to focus their attention on a particular target, a potential source of food or a threat. However, this still leaves unanswered the question of what you might call 'higher' kinds of awareness; in other words, whether animals are really 'conscious' in all of the senses in which that word is used.

## Shared Motor and Spatial Systems

The motor system that was introduced in Chapter 10 on space and motion has its counterparts in all living organisms that exhibit movement. Other species may have bodies that look like our own, and some, like octopuses, will look very different. Either way, they all have some kind of system to control their voluntary movements and trigger their involuntary movements as well. In this mind-based account, we have also been concerned with (human) motor representations, that is, small chunklets of motor knowledge that can be thought of as instructions to various parts of our bodies. They sit there in the motor store with varying resting levels of activation and with associations reaching out to many other types of representations located in other stores, of which the spatial system will certainly be one. Some of these cross-system associations will already be present as part of our total biological inheritance.

## Shared Affective Systems

The affective system, which was already discussed in Chapter 6, has long been associated with the multifunctional 'limbic' system located in what is used to be called the 'old mammalian' brain. This outdated term is still significant, if only in the sense that it signals the fact that at least some parts of our brain are shared with other mammals. Although much research is yet to be done, it would seem that affect as a characteristic of species other than humans is hardly controversial. We can count it as another thing that we share with fellow creatures on this planet, even though the more complex emotions appear to be limited to creatures closer to us in evolutionary terms.

The idea that animals, especially mammals, might feel emotions appears not to be too controversial nowadays: whenever they appear to be afraid or angry or contented and the current situation makes the emotions in question plausible, their outward behaviour is very recognisable from a human perspective. The physical stress or excitement they are experiencing can also be measured. Other species can also recognise emotions in fellow

members of the same species by reading the relevant sensory signals and making the appropriate affective associations. Dogs, for
example, can even learn to recognise the visual and auditory signals of emotions in other dogs and in humans as well.[3] Having,
experiencing and identifying emotions, however, is different from
fully understanding them, that is, being aware of the *nature* of
their emotions. In fact, humans may also on occasions be over-
come by a particular emotion caused, for instance, by grief or
depression and not understand the cause of their feelings and
related behaviour. Nevertheless, it is plausible to assume that our
fellow mammals are also at least 'sentient'. In other words, unlike
a robot or computer program that has been programmed to effectively mimic, that is, reproduce the external signs of given emotions, many animals will be at least aware, without thinking
about it, of their emotional states. Self-awareness in robots is a
popular theme in science fiction, but we have no reason to assume
they have it at present: even though they can mimic human emotions, the apparent self-awareness comes second-hand.[4] Some
non-mammals like birds are more intelligent than was previously
supposed; so, it is very reasonable to place them well ahead of
robots when considering questions of awareness. Bees and ants
with their ability to act intelligently as a group, and other insects
may also possess such awareness, or indeed a group awareness,
although this is more difficult to establish definitively and especially needs a generally agreed notion of how to define 'awareness' in the first place. Human beings, however, will be able to
analyse an emotion, label it and thus provide it with meanings to
reflect on its significance and, if they so choose, to discuss their
emotions with others. Put another way, humans have a highly
*enriched awareness* of their emotions. This is thanks to their
extensively developed conceptual system and their sophisticated
linguistic ability. These two major advantages will now be discussed in turn in Chapters 12 and 13. In sum, it is useful to compare humans and other species since, sharing so much in common
with many of them, we can work out more precisely what the
defining features of the human mind are. The same might be said
of the comparison between artificial intelligence (AI) and human
intelligence, which could be termed 'biological' or 'evolved' intelligence ('EI'). This topic, however, deserves a much fuller treatment than can be given in this particular book.

## Notes

1 De Waal, F. (2017).
2 Nagel, T. (1974).
3 Albuquerque, N. et al. (2016).
4 This line of thinking will be resumed in Chapter 12.

# 12 Conceptualising Consciousness

## Introduction

This chapter discusses the concept of consciousness and how related notions, *awareness*, *attention, subconscious* and *unconscious* can be interpreted within the current perspective. Everyday usage of these terms in different contexts can often be an obstacle in sorting out the exact meanings of these terms. Researchers differ about how much we will ever be able to understand human consciousness. Some claim it is an impossible goal and will remain a mystery forever; others, nowadays, are more optimistic. All these interrelated terms will be integrated within the more general account of activation.

## The Consciousness Challenge

Researchers differ about how much we will ever be able to understand human consciousness. For the time being, and setting aside important debates in philosophy on the subject, there already exists some interesting, relevant research, both in psychology and in neuroscience, containing important insights regarding consciousness concerning both what happens in the *brain* when we are in a state of consciousness and how best to understand this in terms of the *mind*. The aim now is to consider how such insights can be accommodated within the current framework.

DOI: 10.4324/9781003606536-12

## Awareness, Consciousness and Attention

### A Call for Clarity

The discussion now proceeds with the merging of two of the above concepts, 'aware(ness)' and 'conscious(ness)'. However they are used, they actually refer to the same phenomenon. This gets around the problem that, at least in everyday English usage, there are contexts in which one is more appropriate, or more commonly used, than the other. Then, in the current perspective, the actual *experience* of awareness is created by the perceptual systems on the outer ring. This will be discussed in more detail later.

Common usage of words to describe various types of awareness and not only in English can be confusing. For example, you have '*un*conscious' but also various 'dream' states, where the dreamer experiences some form of awareness while asleep. 'Un-conscious', therefore, has a special meaning which is associated with a particular physical state, which may or may not be accompanied by some form of awareness. It is important to be as clear as possible at each point in the discussion about what type or aspect or manifestation of awareness (consciousness) is being talked about and how they are related. One important factor, once activation levels are high enough, is the *degree* of intensity that these levels have attained. The resulting effects on awareness/consciousness are reflected in common usage. For example, you might make comments about your own behaviour, such as: 'I was *vaguely* conscious/aware of something odd going on behind me' as opposed to 'I was *acutely* conscious/aware of it'. Different degrees of activation intensity are what these statements reflect.

### Sensory Awareness

Awareness depends on activation levels being sufficiently high. Apart from the different *degrees* of activation during awareness, it is also important to understand exactly *what* is being activated. While we are awake, we are aware all the time of things happening around us (including, to some extent, in our bodies). This means a) that representations in those systems around the outer ring must be *sufficiently activated* for this to happen and b) the systems must be receiving sensory input. During activation, schemas will be operating that also coactivate outer ring representations. The normal result is continual multisensory experience. Depending on the context, some perceptual representations will dominate others at a given moment, but *all* will be activated to a degree sufficient for awareness. In short, our perceptual portal, the richly interconnected stores on the outer ring are typically buzzing with high levels of activity. They will act as a monitor during our waking hours, producing sufficient degrees of awareness to help us safely navigate our surroundings. This kind of awareness has different labels, and one of them is 'sensory awareness'. We may assume that other species share this particular level of awareness/consciousness. Furthermore, when we fall asleep, as will be explained below, it is not the case that all these activated schemas on the outer ring must all descend into a state of inactivity.

### Being Aware While in an Unconscious State

'Unconscious' is different from '*sub*conscious'. When someone is supposedly in an 'unconscious' state, there may still be awareness, but without the mind's continual updating of what is happening in the environment. It will be a special, limited type of awareness resulting from a physical ('hardware') state, causing one resource, the continued flow of information about what is going on outside, to be inhibited. Without that flow of input, awareness will depend entirely on already stored knowledge being activated and processed, but without any reality checks from outside. This is what happens when dreaming. This means that the perceptual systems on the outer ring, now deprived of the continual updating of what lies outside, are in a sense 'liberated' and operate freely in response to existing *internal* sources of information alone. There is, as it were, a sign for customers on the shop door outside that says 'closed', but inside the shop, there is still activity going on. When an individual *is* dreaming, there may actually be some slight input from the external environment, but it usually gets re-interpreted within the context of the dream:

*A:* 'In my dream, I was aware of being in a dark room with a really uncomfortable feeling of something being wrong'.

B: 'So you didn't realise that your duvet had slipped off and was lying on the floor, right? You weren't feeling chilly?'

A: 'No, indeed. Until I woke up, I was completely unaware of that!'.

Figure 12.1 is a variation of Figure 5.1 in Chapter 5, but with some more detail. It displays three basic states. Separated by horizontal dotted lines, the *resting* state (zero activation) is at the bottom and above that, the middle state involving increasing degrees of activation but *still without awareness* and above that, the conscious state, which begins when the degree of activation has become sufficiently elevated to involve at least minimal awareness. Next to the upwardly pointing arrow is a bar with stripes that do not actually express the idea of discrete levels but simply a continuous increase in activation upwards through all three basic levels. In the case of the middle and upper states, these strips reflect degrees of *activation* rising from minimum to maximum. In the lowest state, where there is *no* activation, they simply reflect differences in resting level. Assume all the time that this is a simplified picture for the purposes of illustration and that the reality is much more fluid: the borders between states will not be so sharply defined.

Keep in mind, finally, that what this picture of mental activation in Figure 12.1 does not capture in any direct sense is the physical state of *un*consciousness. Unconsciousness is a 'hardware'

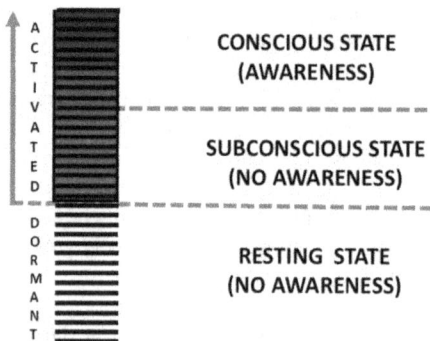

*Figure 12.1* Activation levels, resting levels and awareness.

state that deprives the mind's outer ring systems of any new information from the environment that would otherwise trigger updating of the contents of their five stores together with all their various associations: this deprivation, while it lasts, forces the mind, to rely only on what is already currently stored: the mind happily carries on working with all this previously stored knowledge without any updates.

### Noticing and Attention

The idea of noticing is another concept that needs a more precise explanation. Noticing can, for instance, be intentional or unintentional. There is the intentional, goal-directed type, where something finally catches your eye when you have been deliberately searching for it. Then, there is also the situation when you are only peripherally aware of things going on in the background, the current focus of your attention being elsewhere, or your mind is simply wandering. Something in the background then attracts your attention and forces you to switch away from what you were focused on before. Research indeed suggests that rapid, very minor attentional switches are going on all the time. Some details may remain relatively salient in working memory (WM) but not enough to trigger a switch in attention. Some examples will follow later on.

Whenever we become more alert and attend to something in a more focused manner, our minds become goal-directed. The activation levels of representations related to the object of our attention will increase accordingly relative to others. At this point, the levels may rise up to a point where a more elevated state of consciousness associated with thinking kicks in. The current goal, focusing on specific aspects of what our perceptual systems are currently representing to, becomes more analytic: 'What is that?', 'Why is it there?' and 'What is it doing?'. The engagement of highly activated conceptual representations facilitates more sophisticated responses such as situation analysis, problem-solving and planning. By contrast, at similarly high levels of activation, most fellow species, seemingly lacking this higher level, analytic conceptual ability, will be left to rely on the array of 'instinctive' strategies with which their biological starter sets will have armed them. However, their application of these strategies in particular contexts may have already been further shaped by accumulated associations formed by experience. Examples include learned escape routes and avoidance strategies involving the association of the primal drive to escape harm with specific features of certain familiar situations.

Faced with some sudden and unexpected major threat, humans, like other species, will also react instinctively. We all have a drive, a resident survival goal to control the initial rapid response. The neuroscientist, Joseph Ledoux,[1] tells of the insights that can be gleaned while watching a video replay of the instant responses of people when a bomb explosion had just occurred at the 1996 Atlanta Summer Olympics. The video captures the very first second when everybody momentarily freezes, just as other animals might freeze. At that point, during the early moments of responding to a threat, activation levels will have shot up high enough to allow subsequent reflection in humans. However, where no thinking has yet occurred, the startled antelope and the human respond in a similar fashion, instinctively. For both the human and the antelope, activation levels have shot up to the highest level of intensity. With humans alone, whether exploited or not, higher levels of consciousness at least become available. Depending upon the individual and other circumstances, they may or may not control their immediately subsequent response. Otherwise, both will continue to act instinctively.

## Thoughts as Perceptions

### The Perceptual Engine of Awareness

The role of the outer ring in the explanation of awareness is absolutely central. It is in the nature of all the highly interconnected sensory systems to be activated to high levels all the time we are awake: constant sensory perceptual awareness is required for survival, and we share this with other species. The mind's currently coactivated goals and the processing of the current context will simply determine which particular type of sensory awareness dominates the others at these high levels (the uppermost state in Figure 12.1). This means that the strongly coactivated perceptual systems together form a basic *engine of awareness*. Up to this point, the *objects* of awareness that have been referred to are only whatever has been activated in one or other of the outer ring systems, that is, sensory experiences of the physical environment.

### Sensing Our Thoughts

Since 'the state of activation' is the way in which WM has been defined, another way of describing the engine of awareness will involve the simultaneous combination, potentially with differing levels of activation, of *all* outer ring WMs: somatosensory (body sense) WM, auditory WM, olfactory (smell) WM, visual WM and gustatory (taste) WM. Given the rich interconnectivity of these five perceptual systems and their typically high levels of activation, this special state of awareness has been described as 'global working memory'.[2] This is manifested as a schema of associations that includes representations in all five (sensory) perceptual stores. Under conditions of very intense levels of activation, we can be made aware of something other than our external physical surroundings, something *internal*. This is because associations with representations in outer ring stores also extend into the *inner* ring as well. In particular, they can extend into the *conceptual* store, in other words, implicating the meaning content of our thoughts, things in themselves that are *not perceptual*. Higher-level consciousness somehow happens when these meanings become objects of perception.

Awareness of meaning could be described as a kind of 'sensory projection'. The content of (activated) conceptual representations somehow gets translated into sensory form. Put another way, we sense abstract meaning in a similar way to how we sense our physical environment: concepts, to be experienced consciously, have to become virtual 'percepts'. This explanation is derived from the logic of the activation-based account that has been used so far. It also meshes with the proposals of researchers that the conscious state is characterised by a general, *simultaneous* activation of many types of representation. Moreover, the brain account is similar: intense *synchronised* activity spreads across the whole area of the brain. Stanislas Dehaene even talks of the sudden spread of activation in the brain as a process of 'ignition': neurons suddenly 'erupting' with the onset of conscious processing.[3]

## Projecting the Invisible and the Inaudible

The best we can say about awareness/consciousness at the moment is as follows: despite the fact that *thinking about something* as opposed to *sensing its presence outside* is radically different in nature and easy to grasp, we do see and hear thoughts 'inside' us in essentially the same way that we see and hear things outside us. At least seeing, especially but also hearing, seems to be the best two senses in humans in which to describe the thinking experience. Recall that we can *never* be directly aware of any knowledge representations, especially conceptual ones that underlie our thoughts, nor can we be directly aware of any of our mind's processing operations. We only get the outcomes that can be conveyed to us in a sensory perceptual form. We are, as it were, the 'outside observers' of the results of processing. The only way of (indirectly) understanding what *might* be going on is the kind of analysing, speculating and hypothesising that we indulge in consciously.

One way of imagining how thinking and attention work in this context is Bernard Baars' notion of the 'theatre' of consciousness, with the focus of attention described as a spotlight.[4] Common expressions do express the idea of thinking as (virtual) perceiving. Take, for example, the 'mind's eye' expression, which pictures our thoughts as displayed for us on an inner screen and also statements like 'I *see* what you mean'. However, note that we also say, 'I *hear* what you are saying', which expresses the idea of understanding rather than just literally 'hearing' something. By contrast, we do not say, using less powerful senses, 'I smell what you're saying' (which is what an imaginary dog might do). Perhaps, we could say 'I feel what you are saying', evoking the sense of touch, but here we tend to fall back on a general term like 'sense' as in 'I sense the way you are thinking'. When someone empathises with you and says, 'I feel your pain', conceptual representations will be involved (the 'idea' of pain) but also affective representations (the emotion and its associated negative value).

### Awareness of Emotions

What about becoming aware of our feelings? We certainly become aware of the physical signals that accompany an emotion.[5] In the case of fear or anger, these signals will be especially powerful as our bodies are galvanised into a fight-or-flight state of arousal. In the case of an emotion like sadness, the signals will be quite different. However, as mentioned earlier, probably only humans will have the conceptual resources to enrich our awareness of the emotions and recognise the particular feelings that we happen to be experiencing. Humans are able to analyse and reflect in sophisticated ways on the emotional state in which they find themselves (assuming that they have time to do so): 'I feel so low'. 'I am in a panic: I must try and control my fear'. Sometimes these reflections on a current emotion do not take place, in which case the individual concerned may not yet be fully aware that they are in that state or why they are in that state. They may be seriously depressed but attribute their feelings to the effects of a virus. They may also be aware of experiencing an emotion but cannot attribute any particular cause to it, as in 'Why do I feel so down in the dumps?' or 'Suddenly I feel really happy but I have no idea why!'

### Thinking Happens in Schemas

Thoughts have been described here as based on meanings but not in an isolated sense, that is, not *uniquely* based on conceptual representations alone. The essence of thinking is indeed located in the conceptual system. However, the representation(s) involved in thoughts are part of a schema that will radiate out of the conceptual hub like a spider's web and link up and activate many representations in other stores. The orange schema in Figure 5.1 (back in Chapter 5) provided a highly simplified example of this. 'Imagine a big, juicy orange!': when the thought occurs and the image appears in your mind, activation would have spread from the meaning (conceptual representation) across every branch of the schema, and all the various activated associations with the necessary assistance of the perceptual system will together generate that sensory experience of the thought in your head.

### Effortfulness and Concentration

Energy consumption itself is a physical issue and so more appropriate for discussions about the brain. Nevertheless, the *awareness* of energy being consumed is a mental phenomenon. A state that at times seems to absorb even more energy than heightened attention to something in the outside world, and one which may also induce a distinct sense of effortfulness, is increased *reflective* awareness: this is when we attend closely to the objects of our thoughts. Here, what I still assume to be a uniquely human resource, namely highly co-activated and complex conceptual representations, will play a central role.

Thinking (reflection) is something that can also vary in intensity, ranging from, on the one hand, relaxed, effortless daydreaming, letting thoughts come to us of their own accord at least with minimal direction, to concentrating effortfully on some really difficult problem-solving tasks on the other hand. The intensity reflects the amount of activation that is called upon to carry out such tasks. Concentrated thinking is where that feeling of effortfulness comes in, especially when it lasts a long time and will really burn up the calories.[6] Just imagine trying to solve a tricky bit of mental arithmetic or that frustration of – unsuccessfully – willing some temporarily forgotten name to finally pop up into your head.

All these various states of mind can certainly be sorted out into distinct categories and studied independently, but is there a common thread that unites them all? Part of the answer has in fact already been provided above, namely degrees of (high level) *activation*, the idea of a particular schema or representations *dominating* others and the nature of the current main goal. Some more has still to be said now about how this actually works. Goals will be discussed in a little more detail in Chapter 16.

## Notes

1 See page 6, Ledoux (2002).
2 Truscott (2022).
3 Dehaene (2014) and Dehaene et al. (2006), Dehaene et al. (1998) and also Mashour et al. (2020).
4 See Baars (1997).
5 Damasio called the neural equivalent 'somatic markers' (see, for example, Bechara & Damasio, 2005).
6 Kurzban et al. (2013).

# 13 Different Modes of Awareness

## Introduction

This chapter deals with different 'modes' of awareness, beginning with dreaming, which was already mentioned in Chapter 12. Dream states, occurring while unconscious, might seem to be a problem for defining consciousness. Neither the mind nor the brain is plunged into a completely dormant state. There is activity going on all the time. You can be aware of and think about things happening in the dream world that you have created. The mind draws on all your pre-existing knowledge of the world, all the while, when you are cut off from most of the signals coming in from your immediate external environment. This also happens with regard to less everyday types of states like hallucination, general anaesthesia (sometimes) and hypnosis. Finally, there is what could be called the 'meta' mode. This involves very highly activated conceptual structures and associated representations enabling individuals to be conscious of and fully engaged with the content of their imaginings, feelings and logical thought processes.

### Dreaming and Transitional States Between Being Asleep and Being Awake

The idea that someone can sometimes be 'un-conscious' but at the same time actually conscious (aware) of something is quite confusing because of the words being used to describe different mental states. This needs some clarification. While an individual is alive, the brain is never completely inactive, and the same goes for the mind. During sleep, the brain mechanisms that cause arousal into a waking state are inhibited. Different brain states signal

DOI: 10.4324/9781003606536-13

various types of unconsciousness. While there seems to be little or no obvious reaction to the immediate external environment, even the perceptual stores are by no means completely shut down. Diminished responses to what is going on outside still leave the mind's current perceptual representations in the five outer stores eminently available for participation in processing, together with what is provided in the *inner* ring stores. The rich internal world that has already been created thus far in an individual's waking life can continue to be active, and sometimes highly active during sleep. The mind, being geared to continually make sense of what is going on, will then rely almost exclusively on the current contents of its stores and networks of association.

The relative freedom from information based on input from the external reality is largely responsible for how we experience the dreaming state. We have a rich store of experiences to use for creating any number of alternative realities. While we are dreaming and the outside world is temporarily lost, spontaneous, creative and often bizarre creations of the mind are possible. They are unhampered by the corrective effects of processing and what lies outside us. Our state of awareness is such that it usually excludes any notion of not being awake. Dreams often seem in retrospect to defy logic and common sense. We become absorbed in the internal world that we have created and stored as the representations and their multiple associations that we have accumulated over our lifetime. Awareness then becomes limited to and focused on experiences and thoughts that acquire their own internal logic based on 'impossible' events that now appear perfectly possible. Representational schemas that have been activated do not have to pass any reality checks that, whilst in a waking state, would be vital for optimal negotiation with our environment. The meanings accumulated in the conceptual store are activated, and any unusual associations that seem weird or impossible while awake may be formed during sleeping as part of the general creation of alternatives. These dream phenomena, like ships sailing into harbours and onwards, uninterrupted, over hills and through riverless valleys, would not be discounted as impossible while sleeping, but may be later remembered as weird dreams. Recurring nightmares would be the most dramatic example. Recurrence implies the development of high resting levels following the general principles of activation discussed earlier. This would make them not only more open to recall when the dreamer has woken up but also more likely to be maintained for participation in future dreaming. Dreaming is a very normal state to be in, but sometimes there are less usual experiences like lucid dreaming, where you are asleep but aware of the possibility that you might be dreaming and able to ask yourself the question, 'Is this real or am I dreaming?' There are also other hybrid states where you have become partly aware of your actual surroundings, but with some dreamlike aspects integrated within woven into them.

### Dream Interpretation

Already-formed schemas activated while in a waking state may still be active at low levels during dreaming, especially when they cluster around problems that have claimed our attention while awake. They can then be incorporated and transformed into the versions of reality that are generated when asleep. This makes it tempting to search in remembered dreams for clues that might be meaningful and possibly therapeutic for the individual. Sigmund Freud, the founder of psychoanalysis, was one person who popularised this idea when he was a practising psychoanalyst, and there is a steady flow of books and web-based advice on how to interpret your dreams. Indeed, dream interpretation has been around for millennia and was already taken very seriously in cultures long ago, such as the Sumerian civilisation, well before 2000 BCE.[1]

### Locked-in Syndrome (LIS)

There are unconscious states where it is easy, without evidence to the contrary, for observers to assume the complete absence of any kind of awareness at all. Comatose states may happen spontaneously or may be induced for medical reasons. Comas can last a long time. The crisis point really comes when the brain appears to have lost all but the basic functions that keep the body alive, and some of those may need extra support. At that point, there may still be in the patient some level of awareness of the immediate external environment. In other words, sensory input is still getting through to the outer ring system, and although the whole process is still largely subconscious, there can be enough activation to trigger awareness.

Sometimes, even, a patient is, despite appearances, still sufficiently conscious of what is going on to the extent of understanding what is being talked about. The discovery of this special state
called 'locked-in syndrome' (LIS) was the result of ground-break-
ing research by Adrian Owen and colleagues. It was reported in an
issue of the journal *Science* and published in 2006.[2] When asked to
visualise an activity, for example, playing tennis, imaging (fMRI[3])
of the comatose patient's brain exhibited the presence of activation
in specific areas of the motor cortex associated with performing
the requested activity. The patient in question, a 23-year-old
woman, was therefore more aware of her surroundings even than
she would have been if she had been fast asleep, since, in that case,
such responses to spoken instructions to someone sleeping would
not have been expected. Perhaps, an analogous situation would be
someone in a hypnotic trance, still remaining aware of their surroundings and able to respond to the hypnotist's commands. In
this case, however, the subject would be performing actual movements like raising a hand. In the case of a patient with LIS, if no
voluntary eye movements were observed, any signs of movement
would only show up as related activity in the brain, in which case
the consciousness would be described as 'covert'.

What is interesting in the present context is that it provides a
striking demonstration of how the separate systems in the mind
can collaborate via networks of association, that is, schemas. In
this LIS example, associations would include *linguistic*[4] and *conceptual* representations activated in the conceptual store following the instruction 'imagine you are playing tennis' and *motor*
and *spatial* representations that control the voluntary movements
involved in tennis playing. This suggests to me that more than just
sensory perception of the outside world is taking place because
meanings are being interpreted. When the meaning of the instruction, 'imagine you are playing tennis' pops into your head, motor
representations that you acquired when learning to play tennis
are now activated along with interconnected perceptual representations that will allow you, now the comatose patient, to
picture the requested activity. The activation of the tennis-related
visual representations in the patient now spreads across to all
other associated representations, which are normally activated
when actually performing the physical act of tennis playing. Even

if it were just an uncomprehending response to the sound of the word tennis automatically triggering tennis-related motor representations, it would already signify sensory awareness of the immediate environment. The brain imaging evidence, as well as individual reports of the few people who have recovered, still suggests that there can be an impressive degree of awareness present in the patient responding more as if in a waking state, with movement physically restricted rather than merely awareness of the immediate environment.

There are similarities in this whole process with visualisation-based training. Just imagining that you are practising a skill may have beneficial effects in maintaining or strengthening at least parts of the process involved. Although the physical movements are not engaged or very minimally engaged, the underlying 'programming' is. This includes motor and spatial representations that are associated with the skill in question. To decode the flow of sound of the sentence coming from the doctor (conveying a question to imagine playing a given sport), more is required than straightforward associations between sound patterns and meanings, but grammatical processing as well. There are also implications that follow indirectly from research related to the effects of visualisation. Not only can non-comatose patients with other problems, such as recovering from a stroke, benefit from visualising activity when awake but also people with no such issues at all, such as healthy dancers and athletes, can be shown to improve their physical skills.[5] These are just a few examples of the amazing flexibility and sophistication of the human mind.

## Cogito Ergo Sum

Finally, there is what you might call the 'meta mode' when we are awake and occupied with higher forms of awareness associated with thinking. In this state of meta-awareness, as we are actively occupied with the *content* of our thoughts, we are also at least minimally aware that we are engaged in the *act* of thinking and that its content cannot only be reflected upon but also manipulated and modified. This is the thinking process that convinced Descartes that he actually existed: *cogito ergo sum*, 'I think therefore I am'. Knowledge reflected on and modified that might be labelled 'metacognitive knowledge'. However, it has no separate psychological status as a different *kind* of knowledge except in the sense that it is the product of processing at very high levels of activation. As such, it is unique in humans in that it can be recorded for future use, challenged and changed during exchanges with other individuals and in general has an enormous impact on human life. Without it, complex cultures, ideologies and civilisations with all their physical manifestations could simply not exist.

A common and misleading conception of what the 'mind' is focuses only on that conscious, thinking part. This is, after all, mainly responsible for the knowledge we use when analysing, planning, arguing, designing and imaginatively creating our social structures. It is used for our combined academic understanding and recording of what has been learned and predicted about the universe we live in. It is also used to create the ideologies intended to explain things about our existence that we strongly desire to understand but cannot grasp without their help. Furthermore, we should also add the aesthetic creations to be found in literature, music, architecture and all the other visual arts. This is why, in discussions about the human mind, a type of mental activity that actually accounts for only a tiny proportion of mental activity as a whole gets such a disproportionate amount of attention when discussion focuses on the human mind, rather than the human brain.

### *Effortful Thinking*

The main point about the processing of thought is that, being a conscious activity, it is accompanied by a very high degree of activation. However, even here, there is a range of levels. *Concentrated* thinking requires the highest level, something for which the brain requires a relatively high amount of energy, enough to trigger a sense of effortfulness and often that frown of concentration, facial muscles contracting in a similar way they would respond if we were lifting up a heavy load. This, of course, is quite absent when conceptual representations remain at levels low enough for activation to preclude any awareness of their (nonetheless useful and influential) activity.

### Different Manifestations of the Meta Mode

Our conceptual representations and their schemas adopt new shapes as our thought processes rapidly evolve, millisecond by millisecond. This analytic reflection, so much more than just being minimally aware of something, is a typical feature of *metacognitive processing*. Metacognition is here quite broadly defined as the awareness, as 'outside observers' of our own thought processes, but also of the products of thought, which includes what is commonly called 'knowledge', but also ideas that are disputed. Conceptually based knowledge that is consciously created and manipulated – sometimes also termed *explicit* knowledge – can turn out to be judged unreliable, full of errors or just plain wrong. It can also prove to be reliable or 'more or less' so. It includes *encyclopaedic* knowledge, the general kind which individuals accumulate from reflecting on their experience, but also from outside sources like education and the media. It includes what psychologists refer to as *episodic* knowledge or memory.[6] This is 'life history' knowledge based on episodes in an individual's life and whatever is associated with that, for example, periods or specific points in time when events took place. These may include strong emotional (i.e., affective) associations, in which case the knowledge is labelled 'autobiographical'.

### The Intuitive Mode

Conceptually based knowledge, at least the more complex types of meaning that we can think about, have opinions about, dream about and also communicate to others, appears to be unique to the human species. There is, however, a common distinction between what on a given occasion we *know* is the case and what we 'feel' intuitively to be the case but cannot really *describe* how we know it. Intuitive judgements are best explained in terms of activation levels and the influence of affective factors. Examples of this are provided by Malcolm Gladwell in his book, *Blink*,[7] where he describes people supposedly experts having conflicting intuitions about things which they could not explain. One notable example is experts looking at a small and supposedly highly valuable statue, using their professional (meta)knowledge to assess whether or not it was a fake and some of them having an instant feeling of wrongness while at the same time finding absolutely nothing to justify dismissing it as anything but what it claimed to be. It looked genuine but somehow 'felt' wrong. It turned out ultimately that their intuitions were correct; so, it is likely that their minds had noticed certain details, but without the precision and explicitness and activation levels needed for inclusion in their conscious assessment. These details, however, were nevertheless triggering certain negative associations which together produced a feeling of wrongness in conflict with the expert analysis. The opinion of those with greater expertise was eventually sought in Athens to finally identify the source of these nagging doubts about the statue, and this led to the ultimate verdict that the statue was indeed a fake. Of course, in situations where no conscious evaluation is going on, the intuitive mode, if trusted, can naturally function unhindered.

## Concluding Reflections on Modes of Awareness

### *How did this Happen?*

To conclude the relatively long but still too cursory account of consciousness in these last two chapters, perhaps the most important concern has been to address the question: what causes intensely activated mental representations and their neural equivalents to somehow 'appear' in our heads? Exactly why and how this developed in human minds and brains remains a puzzle, but we have at least made a little headway in an explanation of this elusive phenomenon. As suggested earlier, some cognitive scientists believe a complete explanation is possible, while others see it as a question that will never be fully answered.

### *Is Consciousness Uniquely Human?*

Even though the above account in no way comprehensively explains how consciousness came to be in the first place and precisely how it works.[8] It does perhaps get us a few centimetres closer to understanding consciousness as an experience produced in our biological minds. It also does not, however, explain to what extent and with what precision we can ascribe varying states of awareness to each or any of our fellow species. The French philosopher René Descartes certainly thought that only humans were conscious. And it does seem clear that the advanced conceptual system that humans now possess and the way mental activity can be projected into consciousness place us a very long way ahead of even our closest relatives. Nevertheless, thinking back to the question posed in Chapter 12, it is still difficult not to see this ability in our species and anything other animals can achieve ultimately as a matter of degree: even higher consciousness does not appear to signal a definitive, categorical difference separating us humans from the rest. As established at the beginning of this chapter, 'conscious' and 'aware' are essentially the same thing. They only differ in the way the two words are used in different contexts. So again, no, humans are not the only beings that are conscious. The question is what their states or attainable levels of awareness can allow them to do.

Chapter 14 will deal with human language. This will include the conscious use of the conceptual system to reflect on, control and manipulate language. It will also go into crucial aspects of the acquisition of languages and emphasise the use of the two linguistic systems on the inner ring.

## Notes

1 Black and Green (1992).
2 Owen et al. (2006).
3 fMRI (functional magnetic resonance imaging) measures brain activity as changes in blood flow.
4 To be discussed in Chapter 14.
5 See Ridderinkhof and Brass (2015).
6 In the psychology and neuroscience literature, many of these terms are also used with the word, 'memory.'
7 Gladwell (2005).
8 But see some speculation in the final chapter.

# 14 Communication between Humans

## Introduction

This chapter discusses two main types of communication between humans. Firstly, communication between members of the same species and *without* the use of language takes us to a place where we humans are not so sharply distinguished from our nearest relative. Secondly, when we communicate using language, we are using something that clearly *does* separate us from all other species. Since 'language' and even 'grammar' are terms that have been used when describing the communication systems of other animals, these two terms, along with 'linguistic', will here be restricted to the mode of communication (and private reflection) that is shaped by 'grammatical' structure: here, this means both syntactic and phonological structure. As far as either type of communication is concerned – with or without grammatical structure – 'communication' is about formulating and decoding meanings, that is, conceptual structures. These two alternatives are mutually compatible. In fact, we mostly use both at the same time, communicating with all means at our disposal.

## Defining Two Vital Concepts

This chapter requires that a crucial distinction be made between the more general term, human *communication*, as opposed to human *language*. Communication between members of the same species implies that, one way or another, messages (meanings) are exchanged. There are many ways of doing this, and language is only one of them. So how then is 'language' to be defined? There

DOI: 10.4324/9781003606536-14

are numerous metaphorical uses of this term that have nothing to do with human language. They can include any form of communication between members of a given species as in 'the language of bees' but 'language' is also used in other ways as in, 'body language' or 'the language of the genes'[1] or in AI as, for example, 'large language models', learning algorithms which feed on samples of human language in the public. To present language as something *exclusively* human, we need to narrow our focus and discuss only those crucial, defining elements of language structure which are not found in any other species. This is the focus of Chapter 15. 'Language' will no longer be used to describe just *any* system of communication. It will only be used when the communicative exchange is structured with the contribution of two systems, namely the s and P systems on the inner ring. This also means, for neuroscientists, that what they call 'language networks' in the brain will inevitably involve large areas. To merit the name *language* network, its neural signature must include activity not just from multiple systems, but always activity attributable to what here has been defined as systems, s and P.[2]

## The Conceptual Powerhouse

Chapters 11 and 12 were devoted to the challenging topic of consciousness (awareness). This phenomenon is considered as being not specifically human. Rather, the 'human-only' attribute was applied to the particular mode of consciousness characterised by:

1 very highly activated *conceptual* representations along with any associated and coactivated representations in other stores.
2 the way in which, under the above circumstances, the content of conceptual representations can be projected into *perceptual* form via the combined resources of the systems on the outer ring, giving us a general sense of 'perceiving' our thoughts.

We turn now to communication in general, that is to say, irrespective of what particular resource(s) are used to exchange meanings with fellow humans. Since communication always engages the system where meaning is created and managed, this makes the conceptual system the powerhouse of *all* types of communication. This holds whether or not human language is actively involved in structuring the messages that are passed between people.

It is important to emphasise the fact that humans can still communicate a surprising number of meanings *without* generating a string of (spoken, written or signed) words that has been shaped by grammatical structure. Other animals can communicate with one another by means of body positions, gestures and vocalisations, but without the grammatical resources that the s and p systems together provide. The difference between other animals and humans, apart from these additional two systems that humans possess for communicating, is the sheer number of meanings that humans have available in their conceptual store and the ingenuity with which they can communicate them.

The increase in the range of stored meanings available to humans in their conceptual stores, along with the sophistication of human conceptual processing, increases the potential number and complexity of different messages that could be communicated. This holds even when relying purely on the communicative means that humans and other species have in common. Whenever and for whatever reason, using a shared language is not an option, they will exploit all other resources to try and get their messages across.

### Limitations of Not Using Language

Despite the considerable resources of the human conceptual system, communicating *without* language,[3] that is, 'non-verbally', is still going to be limited when compared with what can be done when using grammatical structure. There are still restrictions on the complexity and conciseness with which meanings can be conveyed between one person to another. For example, imagine trying to quickly signal to a fellow bus passenger without using any words: 'The second newsreader was able to explain the events of the last two weeks more clearly'. The only way to do this without speaking or writing, you might think, would be to use a human 'sign language'. However, if you did, you would, in fact, be inadvertently cheating because this would not qualify as *non*-verbal communication: sign languages have their own grammatical structure and their own visual 'words'.[4]

## Non-verbal Communication

Humans will communicate with one another using the appropriate resources at their disposal, including (but not necessarily) language. This use of multiple resources is the norm wherever possible. Speaking provides the best example of this since, when talking to someone, gestures, facial expressions, so-called 'body language' and various visual aids are used as well. This includes using the voice in ways that are not strictly associated with linguistic structure and which convey emotional states like surprise, disappointment or disgust, for instance. Meanings conveyed in language can be enhanced while speaking by emphasising with gestures (for example) what is being expressed in the form of words. Not only that, but extra layers of meaning can also be added on as well. A facial expression or bodily gesture, for example, can add extra meaning to a simple statement or question by conveying the speaker's attitude to what has just been said. In this way, surprise, irony, scepticism, incredulity and anger can change the whole interpretation of a message expressed in words. A picture or sequence of pictures in a written text can perform the same function. Think also of all the emojis and other icons that are used to communicate in text messages, on screens as well as signs and pictures in many different locations. Think of the visual aids used when teaching, in presentations and in the media generally.

Finally, a point often missed when discussing non-verbal communication is that the conceptual system does allow for different *combinations* of words and non-verbal symbols. Such combinations are manifested in basic communication systems with words ordered solely on conceptual principles, with concepts such as *topic* or *agent* (the doer of an action) or, for example, regularly placing the *agent* of some action in first position ('Samantha' as in 'Samantha hit Bill'. Some claim that these conceptual principles are present in simplified verbal communication, for example, in the very early stages of child language development or in 'pidgin' languages spoken by groups of people who have no language in common. They could be described as precursors to fully fledged grammars, but do not (yet) exhibit the necessary properties that come from processing principles that characterise only *human* language.[5] All this considered, the value and variety that can be achieved with ingenuity, despite the limitations of purely non-verbal communication, should not be underestimated.

## The Origins of Language

At some point in the evolution of the human species, more than 100,000 years ago, human language must have started out as language only in the informal sense, that is, not involving grammatical structure. Without a time machine to transport us back to discover its origins, we cannot possibly know for certain how language really began, but we can at least generate some scientifically plausible scenarios which led to the creation of what we now call 'grammatical structure'. This is an area of research which has sparked a lively debate within linguistics, and this debate has involved looking at our brains and genetic make-up as well as the more easily observable characteristics of human behaviour.[6]

### *The Kanzi Challenge*

What I call the Kanzi challenge is the point where many scientists who are keen to break down the traditional barrier between humans and all other species to prove that the great apes at least also have an ability that we could label 'language'. Looking at how those closest to us, the other primates' terms, communicate with one another does help to show how early humans might have once communicated before language began to take shape. Kanzi was a bonobo, a very sociable, small and highly intelligent ape related to the chimpanzee. He was extensively studied by researcher Sue Savage-Rumbaugh and features in a number of fascinating YouTube videos.[7] His ability to learn how to communicate with a human researcher is extremely impressive. It looks as though his example constitutes a challenge to the idea that language is something only humans can manage. Many watching the videos may well become convinced that Kanzi does have 'language', even in the restricted meaning used in this book. Its importance here, however, is more to do with the light it casts on the way in which early humans and hominids might have communicated before true language ability, as we now know it, began to be formed.

The Kanzi story goes that during the earlier stages of the research project, the young bonobo was watching his adoptive mother learning how to use a display of symbols called lexigrams, each representing some object or idea. Already giving evidence of his lively intelligence, he soon replaced his mother by showing a much greater interest and ability to learn. In the end, he could understand over 3,000 of the lexigrams and communicate messages using a good 500 of them. Many of these messages took the form of not just isolated words but what we might loosely call simple 'precursor' sentences. Note in passing that Kanzi, but not his mother, exhibited this intense curiosity and desire to learn, while *all* human infants with hearing ability will instinctively attend to language sounds (or signs) and end up cracking the inner code of any language to which they are regularly exposed. It is a behaviour that is in our DNA, and therefore in our mind's biological toolkits. Moreover, all children do this instinctively without needing anything like the considerable guidance that Kanzi was given by his human researcher.

### *Speech before Gestures or Vice Versa?*

Although, as we shall make clear shortly, Kanzi had not mastered what makes human language 'language' in the strict sense of the word, but he does give us an idea of what our communicating, pre-language ancestors would already have been capable of (at least in principle, since there were no researchers around with lexigrams and a desire to teach them). Another, and perhaps more instructive, way to investigate the stage when language as we now know it began to take shape in early human society is to look at how modern apes communicate with one another in their natural environment using gestures, other bodily attitudes and movements as well as vocalisations. This suggests to some that speech as we know it may well have been preceded by gestures and simple vocalisations. This is one of the special interests of Cat Hobaiter, a primatologist at St Andrews University in Scotland. Already versed in the fine details of chimpanzee communication with a methodology that had revealed a rich repertoire of ape gestures, she and some colleagues then studied human infants aged between 1 and 2 years interacting spontaneously with each other and their carers. They were able to identify 52 distinct types of gesture.[8] Comparing these infant gestures with the ape gestures that had already been identified, they found as many as 46 of the 52 ape gestures matched those identified in the children. This suggests that as far as communication goes, the ape/human divide cannot be a simple question of 'apes gesture while human speak'. Rather, both species have a shared communicative ability, and they both use similar gestures instinctively, that is, without being taught to do so. The question now is what do they *not* share with humans, and how is it that those meaningful word clusters produced by Kanzi are still missing something that humans have and apes do not yet possess?

At some point in the past, grammatical ability and the systems that it depends on for some reason did begin to develop in early humans as they interacted in their daily lives, not just to issue brief warnings and encouragements but to collaborate more efficiently in hunting, protecting territory, caring for the young and the sick in their group and so on. Several possible explanations suggest themselves. For example, (and speculating now), grammatical development might have been triggered by an expanding conceptual system at a stage where evolutionary pressure built up to create more effective ways of expressing meaning combinations. The emergence of independent grammar system(s) may have stimulated further growth in the conceptual system and, from then on, each system developed into its modern state by a process of synergy, each stimulating growth in the other. As a consequence of this, groups of humans communicated better and faster and were able to build more sophisticated ways of cooperating. In this respect, the sheer importance of having a language was such that the intuitive ability to acquire its grammar eventually became part of the immature child's biological inheritance.

Words expressing particular concepts and sometimes more than one concept are, of course, extremely important for communicating meaning. They have their own internal meaning-altering structure, as in 'un-interesting' (= the opposite of 'necessary'), 'ape-s' (= more than one ape) and 'smooth-ly' (= in a smooth manner). What humans and not chimpanzees can do is not just a question of learning an expanding repertoire of fixed symbols, albeit with a few limited ways of combining them. The real human achievement is about achieving greater communicative power by using a) words[9] and b) complex combinations of words enabling the production and understanding of a *limitless* number of different messages. Toddlers have, without any intention except to communicate meaning, the ability to use the different categories that become attached to words (noun, verb, etc.) and the abstract principles that govern their combinations, which vary from language to language. In this way, without really knowing what they are doing, they succeed in cracking the grammatical code of any language to which they are regularly and frequently exposed. While parents may occasionally teach or correct vocabulary ('no, it's not a horse it's a cow!'), they are rightly confident that the grammar will take care of itself. Which it does, and without any of the worry and effort that some of us feel is necessary when trying to grapple with a new language as adults.

Although many adults do become very fluent in more than one language, it is in practice not always an easy matter to find the right circumstances for achieving what many call a near-native level of performance. An overestimation of the role of conscious rule-learning and the understandable fear of making mistakes are typical obstacles in a formal learning environment. It is much less of a problem experienced by adults living in a multilingual, multicultural community where more than one language is in regular use for the purposes of everyday communication. Whatever linguistic ability adults speaking more than one language may achieve, like children, they will exploit every resource at their disposal to optimise communication with others. Given their maturity, experience and enriched repertoire of meanings, they are likely to be even better communicators than they were as children. And certainly better, I regret to say, than that brilliant bonobo.

## Notes

1 This the title of a book by the biologist, Steve Jones: *The language of the genes*.
2 Forkel and Hagoort (2024).
3 From now on, 'language' will mean *human* language, that is, using s. and p.
4 A sign language is a language equipped, like any other language, with a grammar.
5 Girard-Buttoz et al. (2022).
6 Fitch (2010).
7 https://www.youtube.com/watch?v=ENKinbfgrkU.
8 Hobaiter and Byrne (2014).
9 See also Terrace (2019).

# 15  Human Language Decoded

## Introduction

Linguistic structure, here also referred to as 'grammar', enables the subconscious encoding and decoding of highly complex messages that would not be possible by relying on the organisational resources of the conceptual system alone. What many see as the most miraculous aspect of language resides in the ability of the very young, cognitively immature child to quickly, effortlessly, without reflection and in the absence of grammatical correction, acquire the grammar of any language to which it is regularly exposed. Like other species, humans have their own inherited, instinctive way of communicating what they need and want to communicate. In this way, an individual's total package of thinking and communicating resources is developed to its full potential. This chapter serves as an illustration of how expertise from a given area of research might be used as one way of further elaborating the basic framework. In this case, we look specifically and in more detail at the two systems on which language relies on for thinking and communicating to become characteristically human.

## Elaborating the Basic Account

This chapter serves as one example of how expert systems that create their own representations employ not only *general* processing principles but also an array of their own *special* principles, thereby allowing them to encode their own unique type of simple and complex structures.

The most immediate tasks that are relevant to the present topic are the acquisition and use of natural languages. Inevitably, since

DOI: 10.4324/9781003606536-15

this is an illustration of how the two systems under discussion work in practice, the descriptions will become more detailed: I will try and limit technical aspects to a minimum. Elaborating any part of the framework will require a selection of what is regarded as the best account of the phenomenon under discussion. Accordingly, the further elaboration of this particular part of the framework concerning language could, in principle, be done in different ways. The only proviso here would be that the alternative approach would also need to fit in with the architecture of the mind as described thus far and to abide by its principles.

## The Contribution of Human Language

### *The Core Linguistic Systems*

Despite the considerable contribution of *non*-verbal ways of communicating, nothing takes away from the status of language as a very special human ability, one that, arguably more even than the higher levels of conscious processing discussed in the last chapter, marks us out as different from other species. It proves to be crucial in managing our daily lives from directing two people steering a piano up a winding staircase – arguably achievable without grammar – to tasks impossible *without* grammar such as reading a set of complicated instructions or a contract, not to speak of recording and maintaining historical records, delighting, inspiring and educating us with literature and enabling all the complex social interactions between humans necessary for the maintenance of our civilisation. The extent of the contribution to language communication made by the indispensable two systems, p and s,[1] might look paradoxically small and indirect. This is because, looking at the multiplicity of systems contributing to the acquisition and use of language in its broadest sense, this vital pair represents less than a fifth of the total. That is why 'language' is a far more extensive concept than 'grammar' because it engages multiple systems, most of which have primary functions that are other than enabling communication. These two 'core' linguistic systems shaping language communication are, in this book, referred to as grammatical systems. Note that 'grammar', like 'language', can have broader definitions elsewhere than they have in the current perspective.

Aside from the necessary contributions of P and s,[2] the major player in language development and online processing is, as explained in Chapter 14, the human conceptual system. Trying to detect the meanings in messages expressed in language includes a great deal of conceptual processing. The individual trying to interpret what is being communicated in language will be trying to make sense of it. In the process, actually *two* outside sources will be contributing to the online creation of meaning: firstly, sequences of spoken or written utterances or signing, but secondly, also what else is going on in the situational context in general. When engaged in message comprehension, some part of the total meaning may already be clear or guessable purely from the individual's reading of the situation. This can come across as the mind 'predicting' some of the context of the message in advance of any grammatical processing. For instance, a waiter approaches your table shortly after your arrival at a restaurant, and before any word has been spoken, some associated meanings with what a waiter will usually do and say when arriving at your table will already have been activated and be used in the process of understanding what the waiter *actually* does and says. The individual processing of the current context will automatically activate various associated schemas so that the descriptive term 'predicting' should be taken as a metaphorical description of a purely automatic response to the context without any understanding involved.

## Why Language Acquisition Seems Miraculous

The miracle of consciousness has already been touched upon. There are a number of reasons that one might also want to claim that human language ability is a miracle in equal measure. The first and most obvious reason comes directly from the ideas of Noam Chomsky,[3] who alerted us to the existence of a missing element that logically had to exist given the conditions under which children acquire language. Those one and two year olds gesturing amongst themselves in ways similar to other primates are only at the beginning of the miraculous part of their development. What children go on to is something that modern apes clearly cannot do: they embark on an amazingly fast developmental path without grammatical instruction and without having to reflect on what they are doing, a path that will inevitably conclude with the full mastery of whatever language system underlies the language or languages to which they are regularly exposed. As most of us will know by age 4, children have mastered something that educated and much more knowledgeable adults generally experience great difficulty in achieving, even *with* the supposed benefit of regular instruction and error correction that children do not need. Older learners in optimal learning situations may well achieve very high levels of proficiency in more than one language, but it is *never guaranteed*, as is the case with acquisition in early childhood. In large parts of the world, language acquisition usually means *at least* one language and usually more. Children manage this with no appreciable difficulty, perhaps taking a little more time about it, but assured of ultimate success. Caregivers will concentrate not on children's grammar but on their learning of *words* and their meanings, in this way expanding their vocabulary repertoire. Schooling will help these young children to further develop a) their repertoire of words and expressions, b) more sophisticated ways of using language in different situational contexts and c) more elaborate communicative functions like arguing and persuading. At any rate, by age 4 or even earlier, the basic foundations of the core language system (grammar) are in place.

### Linguistic Versus Meta-grammars

Alongside all the other 'meta' knowledge about ourselves and about the world we live in, we may now add meta-knowledge *about* languages ('metalinguistic' knowledge), that is, knowledge that is developed by thinking about what language is, what languages are, how they work and how important they are. This effectively means that knowledge commonly referred to as 'grammar' can be of two entirely different kinds:

1  Grammars, as previously defined, are built up always subconsciously in the P and S stores.
2  Any knowledge *about* grammar(s) is created as representations in the *conceptual* (C) store.

We must not confuse consciously learned 'metalinguistic' knowledge (in 2) with the hidden structural knowledge (in 1) that is developed in the P and S stores. Following their set of specialised phonological and syntactic principles, the two P and S processors place limits on the possible combinations and associations of their building elements' (primitives) various representatives. They *subconsciously* direct the unreflecting child in its acquisition of its native language(s). Structural solutions that will guarantee the rapid and efficient attainment of a complete grammar are intuitively selected by the child to suit whatever language(s) to which they happen to be exposed. No reflection and no instruction are necessary.

What might be called simple precursor communication systems do emerge, especially in older (second) language learners: a small repertoire of simple words and hardly any variation in word order. Arguably, this type of system has already been seen in chimpanzee communication and is found in adults who suddenly find themselves immersed in a multi-language environment with no common language on which to rely.[4] Otherwise, when acquiring a human language, the P and S systems provide the basic structural options. In addition, in humans at least, it is conceptual principles that often determine how longer stretches of language are organised and which permissible grammatical options are actually chosen in given contexts. In other words, when the core grammatical options have been applied, there is a whole lot of organisational work to be done, especially by the conceptual system, to create an outcome that we can call a recognisable stretch of human language, that is, an utterance (or text) of much more than two or three words.

The P and S systems always operate in such a way that they never require high levels of activation. Conceptual activity, by contrast, can operate both with and without any awareness involved. Being able to rely on subconscious processes is necessary for little children in the very early years of their childhood to be able to achieve the feat of acquiring one or more grammatical systems without ever knowing or even caring what intonation, adjectives or syllables are and what 'wrong word order' implies.

### Learning Grammar Consciously

It should be no surprise that for older children and especially adults trying to learn the grammar of a language, relying at least in part on what they have learned about it consciously, worrying about the rules and, unlike small children, being fearful of making mistakes, should prove difficult, effortful and time-consuming.[5] Thinking itself, especially concentrated thinking, can, as was discussed in Chapter 14, be very effortful: puzzling over correctness and grammatical rules would be an example of this. However, while allowing your *sub*conscious ability to grow, you can still use your metacognitive knowledge, as with other types of skill, to shape your learning experience in a strategic manner in order to get the most benefit from it. How you actually do this might or might not be informed by the expertise of appropriate professionals.

## Cracking the 'Speech' Code

Let us now have a brief look, in this elaborated version of the basic architecture, at what subconsciously 'cracking the code' of the grammar of a given language actually entails. To do this, we have to look at how P and S, the phonological system and the syntactic system, operate. These two systems together constitute what in this book is subconscious 'grammar-building'. What information has to pass into these twin linguistic systems for them to create the appropriate representations for a given language?

The phonological system (P) is conventionally thought of as the system that associates auditory 'sound' representations in the (outer ring) auditory store with its phonological representations, which provides the necessary (in this case, *speech*) structure. Generic sound representations in the auditory system need to be specially structured in P before they can be associated with syntactic structures in S. Readers may be familiar with what speech structures like vowels, consonants and syllables are. These are just a few examples of what phonological primitives will include, with phonological principles dictating all the combination possibilities in a particular language.

### An Example

The auditory representations of an explosion or a creaking door will not produce a coherent response from the phonological processor, which will nevertheless be attempting to make sense of them in its own terms. However, the sound when uttering 'catamaran' will certainly get associated with representations in the phonological system, for example – now putting it in very simple terms, four instances of v(vowel) and five cs (consonants), all arranged in a sequence forming a string of four syllables: *ca-ta-ma-ran*. This combination may be associated in the neighbouring syntactic system with a *noun* (N). However, without any associations reaching from the syntactic system into the *conceptual* store, this string of syllables (etc.) and its current syntactic associations will still be (literally) meaningless. In other words, it needs to get associated with a currently coactivated *conceptual* representation for this particular type of sailing vessel.

For a language to be acquired, subconscious grammatical processing as described above has to take place in collaboration with a host of other systems in the mind that have also been coactivated at the same time. In other words, *multi*systemic schemas will always be involved. Let us take the case of a child acquiring its first language. Picture a room with adults talking sometimes amongst themselves, sometimes to the two-year-old in the room, our language learner. In the same room, there is also a dog, a cat and a parrot (I'll give Kanzi a rest). All living beings in the room are exposed to the same sounds, and each of them is subconsciously representing the sound wave patterns in the manner that their minds and brains have determined for them. For all of them, this means, first and foremost, the activation triggered in their own perceptual systems. For humans, the sounds of speech are first processed, like any other sounds, in the human *auditory* system (AU on the outer ring). Some of these sounds, like the sound of a doorbell, may already have particular meaning associations with representations in the conceptual system (c). However, simple sound/meaning (AU/C) associations in this sense are not enough. As indicated above, a further level of processing (P) is required to build speech structure.

Language is often described as pairing sounds and meanings, but, as already discussed, such associations can easily happen *without* language, in other species as well, and not just bonobos. The dog might recognise the word 'walk', for instance, and prick up its ears in anticipation, whether or not it needs a separate conceptual system to do that. The parrot can memorise sounds and reproduce them, again, whether or not we have to assume a meaning has been formed. So, the sounds that we adults know are language sounds are heard by the child in the room (as well as the dog, the cat and the parrot) as an acoustic stream of sound that is represented and stored in different ways in each of the four listeners.

Stretches of sound generated by such conversations can only become language, that is, words and sentences[6] (and sentence fragments), until at least the phonological (P) system has activated phonological representations to match the activated auditory representations. Recall that auditory representations were based on acoustic patterns detected in the environment with no distinctions made between speech sounds and any other kind of sound. Phonological representations will include representations of stress and intonation, boundaries between words as well as syllables, vowels and consonants. Since the P and S work with anything that comes their way, some of the sounds in the room (like music and doorbells, etc.) will fail to get represented in these two systems.

Adults deliberately communicating with children will supply them with samples of language which they will usually keep simple and comprehensible, guided by the child's responses. Frequently, children will show to their parents' surprise and pleasure that they have advanced further than the parents had assumed. Moreover, caregivers will always focus their attempts on communicating by helping to establish word meanings. One thing they will not be attempting, even if they could manage it, is anything like a series of grammar lessons. They know in advance that the children do not need it. In other words, little children need word learning, and the speed with which their vocabulary grows is amazing. Because the biological starter sets related to grammar building are already in place, and because the children are continually exposed to language utterances, these words will quickly acquire linguistic meaning (minimally) P structure but always some S structure as well. Only then do they count as proper linguistic words, that is, more than simple associations between sounds and meanings.

The direct connections emanating into and out of the twin linguistic systems, P and S, are illustrated in Figure 15.1.

### A Language Based on Smells

Before research started into *sign* languages in earnest using the visual mode of communication, the P system was and is still mostly

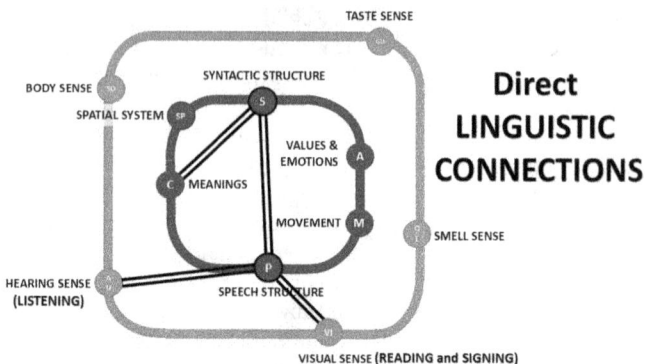

*Figure 15.1* Connections with the two linguistic systems.

associated with the *auditory* system, hence its name: 'P' stands for the 'phonological' from Ancient Greek 'phonee' (φωνή = sound, also voice, and language). Thanks to our fine control of lungs, vocal cords and mouth areas, including tongue, a very large repertoire of sound patterns can also be *produced* by our body, responding to commands from the motor system, for others to understand.[7] In principle, *any* of the senses could be used to express meanings. The question is which of them could then be expressed using the two linguistic systems, that is, as 'language'. In Figure 15.1, the P system connects directly with two perceptual systems on the outer ring (AU – auditory – and VI – visual) allowing it to respond to any activity there, that is, *auditory* activity traditionally associated with phonology (P) as well as *visual* activity which can include both written text as well as signing in a sign language. However, we should also include P responding to Braille characters, hence directly implicating the *somatosensory* (SO) system (a direct connection not shown in Figure 15.1). That said, the tactile input from Braille characters also triggers processing in the otherwise input-deprived *visual* system as well.[8]

Not all of the perceptual systems on the outer ring are equally suitable for supporting language. It is hard to imagine a language entirely expressed as *smell* patterns. For example, just to *interpret* smell messages, we would need something like a dog's highly developed olfactory system to distinguish enough smell patterns to cope with the complex meanings involved in language use, and how would we or a dog *produce* a complex message made up of smells? In any case, since it cannot read or understand language, there is no danger of your dog, if you have one, even attempting this olfactory challenge.

### Signing

The phonological system is, as just indicated, not just about the process of producing and interpreting *speech sounds*. Given the sophistication of human visual ability, many different types of visual patterns can be distinguished. It is not just written text that can be associated with linguistic structures in P and S. Meaningful patterns may take the form of various kinds of bodily (including facial) gestures.

Sign languages must be sharply distinguished from the kind of signing using fingers to spell out messages, or simple signing systems invented by a family at home. Full-fledged sign languages now number over 200. Furthermore, in a given country where, say, Spanish is spoken, a sign language associated with that country will not be a signed version of Spanish but a separate language in its own right.

It seems likely that the P system can also be used for managing visual sign structure as well as speech structure. Human language might have started this way. It follows then that the P system should, as mentioned above, be linked directly not only with the *auditory* system but also to the *visual* system. This would make 'P' seem to have been a premature label to describe this particular linguistic system. However, despite its association with φωνή, that is, sound, sign language researchers seem happy enough to talk of sign language 'phonology', at least for the time being, until a sense-neutral term can be found for P.

### Language in the Written Mode

Writing – also based on visual representations – came along later in the development of human language. It is not something that is naturally acquired by the very young spontaneously, that is, without difficulty, without reflection and without any help. Rather, it is something that gets learnt by older children and adults and not by everybody. The written form of a word accordingly gets first associated with its sound (or, in the case of non-hearing children, the visual equivalent). So, following the normal course of events, assume you have stored the auditory representation of a word like 'cat' in your auditory system, not yet as a proper word yet but as the representation of a sound pattern. Only then can your mind form an association between that auditory representation and the other structures in the phonological store.[9] After that, learning to write can take place. If the sound of the word 'cat' has already become associated with the *meaning* of cat, then this existing sound/meaning association (but not yet a word) is extended by association with *visual* representations corresponding in English to the linear sequence of different visual patterns we call letters, 'c', 'a' and 't'. Association with a corresponding *phonological* representation is then required to make it a full-fledged proper word schema, now including both spoken (and/or signed) *and* written representations of the word 'cat'. In other writing systems, Chinese characters, for instance, different types of associations will be formed.

### Writing Sign Language

In most instances, no commonly used writing system for sign languages has been made available. However, this situation is changing: in fact, as far back as 1974, a dancer, Valerie Sutton, who had developed a notation for dancing called *DanceWriting*, expanded this system to include a notation that is called *SignWriting*, and this has led to the establishment of an *International Sign Writing Alphabet* (ISWA).

### Producing Language and the Motor System

We have been concentrating on listening, reading and also interpreting sign language gestures. As will be already clear, the direction of processing can all go in the reverse direction when *producing* language, except that now we have to, in addition, move various parts of our body (lungs, jaw, lips, tongue, hands, arms, face, etc.) in order to convey the message we have created in our mind. Take speech. If you want to *say* something, you must first form a message in your conceptual system. This will immediately activate associated structures in the syntactic store. Take the dog example. Think of 'dog', that is, activate the dog meaning: its associated syntactic representation, N(OUN) or N(OUN) P(HRASE), will immediately be coactivated, and this in turn will co-activate associated phonological representations. These, in turn, will co-activate associated sound representations in the auditory system. Finally, in order for speech to be physically created to convey the intended message, the appropriate muscles have to receive commands to behave in certain ways. Since the auditory system has a direct link to the motor system (on the inner ring), as shown in Figure 10.1, its auditory representation can be associated with *motor* (M) representations in the motor store. There is also a direct connection between the motor and the *spatial* system (again shown in Figure 10.2 but not included in Figure 15.1), which will also be a vital component for the whole operation of getting the body to produce the right movements in order to articulate the word 'dog' in the speech mode.

In writing or in sign language, it is basically the same process as above, except that *visual* representations of the patterns to be created in written or sign language form would be the most important ones. Since spreading activation will trigger all kinds of associations, some of them will always be currently more highly valued than others. If the context decrees that *speech* is the intended mode of communication, the *auditory* representations will be boosted far above any coactivated visual ones. And vice versa.

The final result as regards the mode of message production is concerned will depend upon which, in the whole activated schema, are the most highly activated ones. If the context requires auditory representations, they, along with their particular associations in the motor and spatial stores, will all get that activation boost needed to win the competition. You will then *say* rather than write 'dog' by articulating your organs of speech, completely unaware of all the complex and rapid underground activity going on to make it happen. It is highly unlikely that your hands will ever betray even the slightest twitch in response to the weakly activated visual representations in the dog schema that would have been required for writing the word 'dog'. If, however, the context includes the intention to *write* the word, then the motor responses associated with speech production will be minimal, whereas associated *visual*

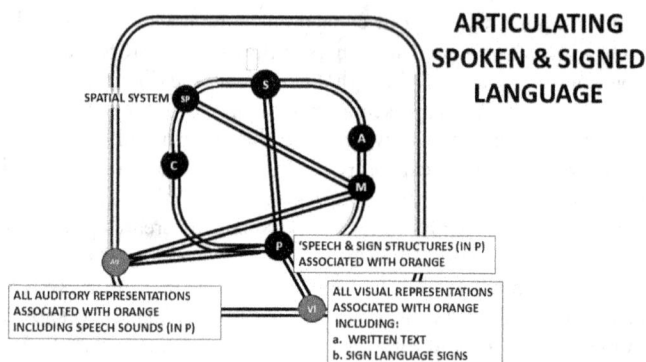

*Figure 15.2*  The articulation of a spoken or signed word: the example of 'orange'.

structures (written signs) will be *strongly* activated and triggering motor responses associated with writing and/or typing will dominate those associated with speaking. For those readers curious about how this would look on the metro map, Figure 15.2 is a schema fragment that, apart from the essential connections between the spatial and motor systems and those linking the syntactic and phonological systems, details only the required connections for speaking or signing a word (in this case 'orange').

### Linguistic Primitives

Identifying linguistic primitives in the P and S systems has been an extensively researched area, although the details will differ depending on the theoretical approach. In terms of this book, preference in the current elaboration is given to linguists convinced that the regular recreation of a language system in each immature child's mind could not possibly be explained by the human conceptual system alone. Just as with all the other primitives that drive each and every one of the mind's systems presented in this book, we rely on the researchers in a given area to develop plausible accounts for what the two types of linguistic primitives – the P primitives and the S primitives – might look like. What this really means is that any language is acquired and used by engaging multiple systems and therefore many types of primitives. The ones that make it a 'language' rather than just any system of communication are the organising principles and primitives of these two above-mentioned systems, the one that takes care of phonology and the other that takes care of syntactic structure as described in this chapter.

## *The Continuing Contribution of the Conceptual System to Language*

As a final acknowledgement of the multisystemic nature of human language, the contribution of the conceptual system is owed the final spot. It contributes not only 1) meanings in general, 2) a default communication system in special situations where language is not a possible mode of communication aspects and 3) an auxiliary system controlling non-verbal aspects (signs, gestures, etc.), when the P and S systems are active during communicative exchanges, but also contributes to shaping of language in response to different communicate intentions and contexts of use and to creating longer coherent and cohesive utterances beyond the single sentence. This covers a large area of language research associated with various label terms, notably *discourse* and *pragmatics*.

From the very early stages of an individual's development, the meaning system (C) gets involved in communication between humans. Even before the child starts to speak, it facilitates the exchange of meanings. It accompanies the acquisition of a language's grammatical system and ends up making its major contribution throughout the educational years as the child/adolescent/adult learns how to use the language effectively. All of these contributions, by definition, are provided via the application of the construction principles, not of the P and S systems, but the C system.

## Notes

1 Mainstream generative linguistic research thus far has been carried out by identifying just one grammatical system, syntax, while this book follows a twin system account following (Jackendoff, 1997).
2 Although the framework account is not identical in every respect to Jackendoff's ideas, it nevertheless owes a big debt to his explanation of language and his insistence on integrating explanations of language with the explanations of the mind as a whole.

3 Chomsky (1965, 1986).
4 See, for example, Klein and Perdue (1997).
5 Certain schools of thought recommend a wholesale concentration on learning to communicate without explicit reference to grammar at all or with only occasional and brief reference to rules. The no. 1 requirement is lots of regular exposure to the language in situations where most of it is comprehensible or rendered comprehensible one way or another.
6 For humans, written text, also in Braille, require a related type of explanation.
7 Sutton (2009).
8 Haupt et al. (2024).
9 The reader is spared the more technical detail of these phonological primitives.

# 16 Concluding Reflections

## Introduction

Chapter 16 briefly reviews all the previous chapters taken as a whole. Then, there will be a brief discussion of some additional topics which, due to space limitations, could not be fully covered earlier. Some, more speculative reflections follow that are intended to stimulate more thinking about the mind, its relationship with the brain and implications for ways in which we humans use the power of thought to cope with what we cannot yet fully understand.

## Summing Up

The preceding chapters provided a means of looking at the mind from a particular perspective. One aim has been to present the mind and brain as the joint biological product of evolution. This reflects views that were common even when cognitive science was in its infancy. As mentioned in the preface, to explain the mind in a coordinated and comprehensible manner, a particular perspective to guide this process was chosen, namely the Modular Cognition Framework. Although the number of mental systems involved has been restricted to 11,[1] they could, in principle, be broken down further by adding more modular systems, to account for, say, logical, mathematical and musical ability. Our reluctance to embrace massive modularity determined the choice of what seemed to be the most parsimonious of plausible options. Using this framework, it becomes possible to present the mind

DOI: 10.4324/9781003606536-16

coherently and succinctly by first outlining its relatively simple architecture, its general processing principles and the additional principles defining each system and used to build and organise its own representations.

Despite being differently coded and hence mutually incompatible, each system's representations can still be associated with and hence coactivate other types of representations. Systems can thus collaborate in different ways to solve the steady stream of tasks that life throws at us. They operate at subconscious levels, with conscious processing forming an exceedingly small part of the whole. Furthermore, the status of the mind as *biological* software had the advantage of permitting the use of what might be an otherwise misleading metaphor. It also describes the mind's equivalent of DNA in terms that highlight the exceptional contribution of what humans inherit at birth. The software analogy avoids misinterpretation by virtue of its association with biologically 'evolved intelligence' (EI) in clear contrast to its impressive but nonetheless much inferior artificial counterpart, AI. Taken together, the brain and mind sides of the coin constitute the essence of what it is to be a human, a story with two complementary ways of being told.

These days, questions about human cognition often get investigated in terms of the physical brain using all the rapidly developing technology for combing through its neural structure. Perhaps, amidst the excitement about all this progress in neuroscience, we can get a little too starstruck by the constant flow in the media revealing yet another supposedly major breakthrough in brain research. This book should also act as a reminder, if one is needed, that we can still, if we so wish, talk about the biological *mind* independently *without any obligation to discuss neural structure and neural activity*. That said, for a full account of who and what humans are, neuroscience and psychology will always need each other.

This closing chapter now continues with a number of relevant issues that were not given enough space in the preceding chapters and would probably each deserve a separate book-length treatment. Accordingly, the following topics will only be dealt with briefly by way of an incentive to stimulate further thinking and debate.

## Complexity Versus Simplicity in the Mind

The account of the mind's architecture as presented in this book is at the very least a scientifically plausible one. The family of systems arranged around two rings is an idea that should be easy enough to grasp and provide a convenient way of introducing the essential features of the mind's organisation as opposed to its neural manifestation. This relatively simple and straightforward set-up is actually a very powerful one. Paradoxically, it endows the mind with the capacity to handle much of the enormous complexity of human experience. It does this with admirable flexibility and efficiency. This is precisely what it has evolved to do.

The most salient characteristic of the mind, one that already stands out in the metro map, is functional *specialisation*. Each contributing system responds to *its* part of any complex task that comes along by using its particular internal principles swiftly, efficiently and in parallel with others. Functional specialisation can thus be best described as a *biological* solution to the complexity, variability and fluidity of life experience. What actually happens in the mind with its dedicated systems and stores is accounted for in terms of *activation*: this is the chosen major source of explanation for how the mind actually *works*. More gets activated that will actually ultimately be used for a given solution, so a key factor in reaching solutions is *competition*, where the currently most highly activated knowledge representations have the upper hand (following the 'law of the jungle'). Changes in the current situation affect the relative dominance of particular representations and schemas by bringing about shifts in priorities. These priorities are determined by whatever happens to be the current context at any given moment. What is a winner one moment can quickly become a loser, and vice versa. Importantly, 'context' here is what gets to be processed and interpreted by the mind at that particular moment, and not what a body of observers might call the 'objective' situation. It includes not only what ongoing states and events are being represented by the mind's outer ring systems but also the contribution of activity in the inner ring systems, adding meaning, emotion and value to name but three sources of influence emanating from the inner ring.

Another aspect of the mind's operations is the fact that the systems work in parallel and are not limited to some sequence, which has already been mentioned. Also mentioned was the fact that the mind does *not* work as a strict hierarchy of systems with one controlling system at the top. Even though hierarchical relationships will still be present in many operations, the mind is best described as fundamentally *heterarchical*. This means essentially that the locus of control is not fixed but shifts according to circumstances as currently interpreted by the mind. Heterarchy is another biological solution to resolving the complexity and dynamic flow of states and events in the outside world. These guiding principles govern how we cope with reality and reduce its complexity. They are essential for a proper understanding of how the mind works.

## Goals

### Goals as Intentions

One of many topics that could not be fully developed was 'goal', and another related one was the notion of 'self'. They are the subject of much ongoing research in the different branches of cognitive science. Goals have certainly featured in several chapters and are most commonly understood metacognitively, that is, as the product of *conscious* planning, of a deliberate intention to achieve some result: 'I intend to buy a cheesecake' or 'I want to walk home now'. As such, they are first formulated consciously in the terms of the conceptual system and elaborated, mostly subconsciously, as a set of schemas that are dominated by the main goal and a series of subgoals that serve that main goal.[2] Goal-based schemas will also inevitably involve the coactivation of representations from many different stores.

Some goal-based schemas or parts of them will have been frequently used and no longer require much in the way of attentional resources to execute the tasks for which they were created. Once the main goal has been decided, walking home, for example, may require only a minor degree of awareness. There may be no need for any high-level planning of the route, just the business of avoiding predictable minor obstacles on the way like other walkers or roads to be crossed. Goals based on novel situations may demand some hard thinking and can therefore become effortful. One example would be a main goal like obtaining a long list of diverse items in an unfamiliar town, requiring a host of sub-goals involving checking the time, the weather and the availability of suitable sources.

### Goals as Drives

Apart from deliberately conceived goals, there are also *inherited* goal schemas that have their origins in basic biological drives. These are triggered when there exists, in some area such as hunger, thirst, sex, security and curiosity, a current state of tension and imbalance with the corresponding need to rectify it. Drives are variously described and categorised in the psychology literature and were briefly discussed in Chapter 8 in relation to the current framework.

Subconscious drives and the ways they operate can also be described using terminology applied to consciously conceived goals with terms like *decision, selection, executive, strategy* and *prediction*. It is important to understand that, in this case, these are metaphors and not to fall into the trap of thinking there is some hidden mind-within-a-mind, a hidden *homunculus*, planning and directing operations. Affective representations, especially values, will be part of any goal-based schema.

## Self

### What and Who You Think You Are, or Else What?

The bases of self are often viewed as a coalition of different selves. There is certainly a set of meanings and associations making up a complex 'ME' concept expressing who you are and what you are. The main value of a ME-dominated schema, your 'meta-self', is not so much that it reflects the totality of who you really are but rather your *conception* of it: your actual behaviour can often run counter to your idea and expectations of yourself. One thing the meta-self, the conscious 'you', does is act as a constant reminder of your separateness as an individual, different from other people. This is a particular awareness of selfhood that is, perhaps misleadingly, referred to as your 'theory of mind' (ToM), an awareness which begins to emerge in early childhood that you are not the only self and there are other minds and view-points that require your attention. Because the 'you' that is expressed in the meta-self is something that you can think about, worry about, be happy with, talk about to others in different ways, hide from others on occasions and which you use to explain to yourself, it will be familiar to most of us. It will involve conceptual-affective associations, reflecting how you feel about and critically evaluate yourself. However, it is also worth remembering that the seeds of selfhood are already present in the somatosensory system in the form of proprioception: this is the 'sensed self' and includes a sense of your body, its position and balance. *Interoception* can also be seen as an aspect of the somatosensory system: it gives a sense of what is going on physically *inside* your body. From a mind point of view, although this involves signals coming from inside your body, those signals are nevertheless part of the *external* world that the mind needs to interpret. The resulting representations will therefore get associated with both positive and negative representations in the *affective* store, along with many meanings in the *conceptual* store. Perhaps, the real self is not really what you think it is, that is, your meta-self. Rather, it is best thought of as the totality of your inherited toolkits in each system. It is all the predispositions that you had at birth as an individual *plus*, importantly, all the subsequent modifications brought about by your life experience to date.

## The Scope of 'Cognition'

The chapters of this book each reinforce the notion that 'cognition' and 'the mind' include both conscious and especially subconscious aspects, the latter accounting for most mind (and brain) activity, the former, however, capturing much more attention for a number of reasons. One reason for this is that higher-level awareness (thinking) depends upon conscious processing. It makes this aspect of cognition extremely useful for the purposes of communication, recording, and, in general, creating cultures and civilisations, organising the external environment in complex ways to satisfy the combined needs of individuals and groups. The result is that using this higher level of consciousness, we create, we philosophise and we satisfy in various ways our curiosity about the world we find ourselves in and control it as much as we can with varying results both good and bad. All this, of course, is very much due to the power of the human conceptual system and also the resources provided by human language. Apart from the high value placed on metacognition for what it seems to have done for us as individuals, it remains difficult – but important – to consider seriously and think clearly about all that is hidden from higher-level awareness, including all the influences that we cannot easily control.

## Mind over Matter

Since the mind as presented in this book is a biological phenomenon, the issue of mind over matter might seem just a little easier to solve, but science is still a long way from uncovering the actual processes that cause people to be affected by, say, the power of suggestion. However, consider how people react to threatening, stressful events that are not ascribable to the body's automatic reactions to physical trauma. It seems obvious that the response of the brain's amygdala, triggering the release of neurotransmitters such as cortisol and adrenaline, is a brain-based response to stressful situations. Sometimes the interpretation of threat is just a hardwired physical reaction, like the momentary freezing response to a very loud sound. In other situations, you could say that the mind is *interpreting* a given situation as threatening while its biological hardware is providing the physiological response. Note that when making such separate statements about mind and brain, the background assumption can still be maintained, namely that the two form an integrated whole.

The most intriguing question is not whether the mind (together with the brain) *can* affect matter, but more of how much in general and especially *in what circumstances*. There is, for example, evidence for the reality of *placebo* and *nocebo* effects, indicating that sick people's beliefs that they are going to get better or, in the case of a nocebo, get much worse, even die, can sometimes have these self-predicted outcomes.[3] There is also evidence for limitations on mind-over-matter effects, as when the conviction that some serious disease can be halted and reversed by the power of thought (including prayer) does not produce the desired outcome.[4] Deliberately conceived strategies to directly influence the course and outcome of an illness may often fail. If they appear to work, believers rarely double-check the history of outcomes that might explain recovery because they already have their own explanation.

Without knowing more about the mechanisms involved in these types of situations, it is still difficult to see why sometimes things work and sometimes do not. These phenomena do at least fit in well with the idea that the mind and the brain are part of the same biological story. Intentions and ideas in general are, of course, abstract concepts. You can't see or touch them, but we accept their reality. To the extent that we are aware of them, they can be treated as products of the *mind*. And we, as humans capable of this high level of consciousness, can be *aware* of our intentions and ideas, but the actual processes that have activated them are subconscious and therefore beyond awareness. We can assume that ideas will always have their neural manifestations. These will take the form of particular, complex patterns of electrochemical activity (albeit currently a big challenge to properly identify).

Intentions can be consciously formed and modified irrespective of any hidden drives involved. They can also be communicated to other people. Once they are processed, recreated and stored in those other individuals' minds, they will necessarily have some physical manifestation in their brains as well. Such communicated intentions, as ideas stored in other minds and participating in their conscious thought processes, can, at the same time, exert observable physical effects in those other individuals' bodies. Some of those influences might appear miraculous. However, mysteries such as answers to a deeply felt wish or prayer and including nocebo or placebo effects, need not be attributed to some type of paranormal transference between individuals, wherever language and other types of communication between humans could have done that job perfectly well.

## Free Will

If complete free will is an illusion, it will not be at all apparent to us as we plan ahead and make conscious decisions. This is because when reflecting on this notion, in my meta mode, it does appear to me that 'I '(my ME-dominated schema) has the power to plan and make decisions on which possible course of action to take in any given circumstance. I can also freely make value judgements about the degree to which I achieved what I wanted and whether it was worthwhile, worthy or unworthy. Most animals lacking the ability, at least on the face of it, to pause and reflect on the choice between different possible courses of action would seem to rely on instinct. Where there is a conflict – will they fight or will they flee? – they will act according to their instinctive assessment of the situation and dominant impulse at the time. However, given what has been said about hidden, subconscious influences that influence the conscious evaluation of one option over another, we humans are forced to conclude that *complete* free will, in the sense of a *consciously* constructed decision that is free of any hidden internal influences beyond our control to make and implement some chosen course of action, might be extremely difficult to attain if not completely illusory. It must probably be a somewhat limited version of what we would ideally like it to be. We should be glad, as humans, of having at least a sufficient level of control over our thoughts and actions to feel and be able to claim responsibility for them.

## The Big Questions about Life

### Transcendental Answers

For those who believe that there is another dimension of reality, that the mind survives in some form after death and that this is a part of some greater plan, there are particular belief systems that provide a basis for some or all of these ideas and answers to how life should be lived accordingly. Since they are all related to what are called spiritual matters and since these 'transcend' the laws of nature as scientists currently understand them, then assessing their truth value is by definition beyond the scope of this book. That said, it is still possible to provide suggestions in terms of the biological mind as set out in the preceding chapters about how and why these ideas and belief systems came into being. This ties in closely with what was said earlier in the sections on goals and self.

The biblical myth of what drove Adam and Eve from the Garden of Eden – eating an apple from the Tree of Knowledge – and, in a different way, like the story in Greek mythology of Pandora's box, from which, once opened, all the evils of the world flew out together – with just one thing – hope – left behind – can provide clues about the origin of transcendental thinking. The idea behind such myths seems to reflect the big life questions that thinking people ask themselves and passionately want to know the answer to. What are we here? What is the meaning of life, or is there no meaning independent of what we provide ourselves? Why is there so much violence and destruction, human-made or otherwise, with so many innocent people suffering? Why are some things so incredibly beautiful?

   The responses to such questions can be interpreted as the imaginative creations of human thought in response to deep-seated drives (goals) for security, social bonding and curiosity. Because of the development in humans of higher consciousness, this goes well beyond working out the best way of surviving and thriving in the local environment. The development of higher consciousness has given humans access to painful questions about life they would never otherwise have asked. Higher consciousness – represented here as Eve's apple, the forbidden fruit of the Tree of Knowledge, proves to be a very mixed blessing. Humanity is no longer free to roam like innocent and ignorant children in the Garden of Eden. Once the apple has been eaten, there is no going back. Big Questions, in other words, beg for Big Answers. Scientific explanations, being non-forthcoming or emotionally unrewarding, answers are sought elsewhere.

   When seeking ways to optimise their survival and thrive on this planet, more thoughtful humans may experience the agony of ignorance when contemplating the possible meaninglessness of existence. For many, even the most intelligent among us, this is not a tolerable state in which to remain. The sought-after solution is imaginatively created and, in some cases, is provided ready-made in institutionalised forms in order to ensure a feeling of meaningfulness, self-worth, comfort, security and structure in life. That is a big advantage. The downside, of course, is that any ideology, political, religious or otherwise, can be and too frequently abused. It is indeed an unsettling thought that if all things remain meaningless unless we humans invest them with our own *created* meaning, the extinction of our species on this planet might just have a devastating consequence. As the physicist Brian Cox has suggested, if we are the only ones that can think in any of the two trillion galaxies that make up our universe, all meaning, including the meaning of life itself, would be permanently extinguished.

Understanding how the mind works and how those inherited goals – the drives – can motivate us to conceptualise a world outside us that can satisfy those drives may be one way of understanding why some attempts to make life meaningful are harmful and lead to chaos and destruction, while others are helpful and benefit the common good. It would seem to depend on which particular drive is the dominant one. All things considered, then, in our current stage of evolution, emotion (the affective system) is still in the driving seat. The human capacity for reason seems sometimes to be relatively better at aiding and abetting our desires rather than controlling them. But then again, there is always what Pandora prevented from flying out of the box, hope. This is why there are still those who go on trying to create both rationally and spiritually inspired solutions in order to create a better world for us to live in.

## Notes

1 This may be compared with an otherwise similar approach that espouse 'massive modularity' (Carruthers, 2006).
2 For further discussion of goals, see Truscott and Sharwood Smith (2019), pp. 174ff.
3 Goldacre (2008).
4 Benedict (2006).

# Bibliography

Albuquerque, N., Guo, K., Wilkinson, A., Savalli, C., Otta, E., & Mills, D. (2016). Dogs recognize dog and human emotions. *Biology Letters, 12*, 20150883. https://doi.org/10.1098/rsbl.2015.0883

Asimov, I. (1942). Runaround. *Astounding Science Fiction, 29*(1), 94.

Baars, B. J. (1997). *The theater of consciousness: The workspace of the mind*. Oxford: Oxford University Press.

Barrett, L. F. (2017). *How emotions are made: The secret life of the brain*. London: Macmillan.

Bechara, A., & Damasio, A. R. (2005). The somatic marker hypothesis: A neural theory of economic decision. *Games and Economic Behavior, 52*(2), 336–372. https://doi.org/10.1016/j.geb.2004.06.010

Black, J., & Green, A. (1992). *Gods, demons and symbols of ancient Mesopotamia: An illustrated dictionary*. University of Texas Press. ISBN 10:0292707940.

Boomer, D. S., & Laver, J. D. M. (1968). Slips of the tongue. *British Journal of Disorders of. Communication, 3*, 2–12.

Carey, B. (31 March 2006). Long-awaited medical study questions the power of prayer. *The New York Times*. Archived from the original on 12 March 2013.

Carruthers, P. (2006). *The architecture of the mind: Massive modularity and the flexibility of thought*. Oxford: Clarendon Press/Oxford University Press. https://doi.org/10.1093/acprof:oso/9780199207077.001.0001

Chomsky, N. (1965). *Aspects of the theory of syntax*. Cambridge, MA: MIT Press.

Chomsky, N. (1986). *Knowledge of language: Its nature, origin, and use*. New York: Praeger.

Cohn, N. (2013). *The visual language of comics: Introduction to the structure and cognition of sequential images*. London, UK: Bloomsbury. 240. ISBN 9781441181459.

De Bot, K., & Clyne, M. (1994). 16-year longitudinal study of language attrition in Dutch immigrants in Australia. *Journal of Multilingual and Multicultural Development, 15*, 17–28.

De Waal, F. (2017). *Are we smart enough to know how smart animals are?* New York: Norton.

Dehaene, S. (2014). *Consciousness and the brain: Deciphering how the brain codes our thoughts.* Viking.

Dehaene, S., Changeux, J., Naccache, L., Sackur, J., & Sergent, C. (2006). Conscious, preconscious, and subliminal processing, *10*(5), 205–211.

Dehaene, S., Kerszberg, M., & Changeux, J. P. (1998). A neuronal model of a global workspace in effortful cognitive tasks. *Proceedings of the National Academy of Sciences, USA, 95*, 14529–14534.

Dodier, O., Barzykowski, K., & Souchay, C. (2013). Recovered memories of trauma as a special (or not so special) form of involuntary autobiographical memories, *Frontiers in Psychology*, *14*, 1268757. https://doi.org/10.3389/fpsyg.2023.1268757

Eckman, P. (2007). *Emotions revealed, second edition: Recognizing faces and feelings to improve communication and emotional life.* New York: Owl Books.

Fitch, W. (2010). *The evolution of language.* Cambridge: Cambridge University Press.

Forkel, S., & Hagoort, P. (2024). Redefining language networks: Connectivity beyond localised regions. *Brain Structure and Function*, *229*(9), 2073–2078. https://doi.org/10.1007/s00429-024-02859-4

Freud, S. (1997). *The interpretation of dreams* (A. A. Brill, Trans.). Wordsworth Editions. ISBN, 1853264849, 9781853264849.

Girard-Buttoz, C., Zaccarella, E., Bortolato, T., Frederici, A., Wittig, R., & Crockford, C., et al. (2022). Chimpanzees produce diverse vocal sequences with ordered and recombinatorial properties. *Communications Biology*, *5*(410), 1–15. https://doi.org/10.1038/s42003-022-03350-8

Gladwell, M. (2005). *Blink: The power of thinking without thinking.* New York: Little, Brown and Co.

Gods, demons and symbols of ancient Mesopotamia: An illustrated dictionary. Austin, Texas, *Oneiromancy.* Wikipedia. Date retrieved: 21 April 2025 10: 36 UTC https://en.wikipedia.org/w/index.php?title=Oneiromancy&oldid=1264682612

Goldacre, B. (2008). *Bad science.* London: Fourth Estate Ltd.

Handley, I., Brown, W. E., Moss-Racusin, C., & Smith, J. (2015). Quality of evidence revealing subtle gender biases in science is in the eye of the beholder. *Proceedings of the National Academy of Science USA, 112*(43), 13201–13206. https://doi.org/10.1073/pnas.1510649112

Haupt, M., Graumann, M., Teng, S., Kaltenbach, C., & Cichy, R. (2024). The transformation of sensory to perceptual braille letter representations in the visually deprived brain. *eLife*, *13*, RP98148. https://doi.org/10.7554/eLife.98148.3

Hobaiter, C., & Byrne, R. W. (2014). The meanings of chimpanzee gestures. *Current Biology*, 24(14), 1596–1600.

Ishikawa, T. (2021). *Human spatial cognition and experience: Mind in the world, world in the mind*. New York: Routledge.

Jackendoff, R. (1990). *Semantic structures*. Cambridge, MA: MIT Press.

Jackendoff, R. (1997). *The architecture of the language faculty*. Cambridge, MA: MIT Press.

James, W. (1890). *The principles of psychology*. New York: Holt.

Javadi, A. H., Emo, B., Howard, L., Zisch, F., Yu, Y., Knight, R., Pinelo Silva, J., & Spiers, H. J. (2017). Hippocampal and prefrontal processing of network topology to simulate the future. *Nature Communications*, *8*, 14652. https://doi.org/10.1038/ncomms14652

Jones, S. (1994). *The language of the genes*. London: Flamingo.

Kahneman, D. (2011). *Thinking, fast and slow*. London: Penguin Books.

Kandel, E. (2012). *The age of insight: The quest to understand the unconscious in art, mind, and brain, from Vienna 1900 to the present*. New York: Random House.

Kanzi: The Ape that Understands Humans and Knows Over 3000 Words. https://www.youtube.com/watch?v=ENKinbfgrkU. Accessed on 17 March, 2015.

Klein, W., & Perdue, C. (1997). The basic variety (or: Couldn't natural languages be much simpler?). *Second Language Research*, *13*, 301–347.

Kolibius, L. D., Josselyn, S. J., & Hanslmayr, S. (online, 2025). And yet, the hippocampus codes conjunctively. *Trends in Cognitive Science*, *2738*. https://doi.org/10.1016/j.tics.2025.06.013

Kurzban, R., Duckworth, A., Kable, J., & Myers, J. (2013). An opportunity cost model of subjective effort and task performance. *Behavioral and Brain Sciences*, *36*, 661–679. PMID: 24304775; PMCID: PMC3856320.

Langer, E. J. (2009). *Counter clockwise: Mindful health and the power of possibility*. New York: Ballantine Books.

LeDoux, J. E. (1996). *The emotional brain: The mysterious underpinnings of emotional life*. New York: Simon & Schuster.

LeDoux, J. E. (2002). *The synaptic self*. London: Penguin Books.

Loftus, E. (1979). *Eyewitness testimony*. Cambridge, MA: Harvard University Press.

Maguire, E., Gadian, D., Johnsrude, I., Good, C., Ashburner, J., Frackowiak, R., & Frith, C. (2000). Navigation-related structural change in the hippocampi of taxi drivers. *Proceedings of the National Academy of Sciences*, 97(8), 4398–4403. https://doi.org/10.1073/pnas.070039597

Mashour, G., Roelfsma, P., Changeux, J-P., & Dehaene, S. (2020). Conscious processing and the global workspace hypothesis. *Neuron*, 105, 776–778. https://doi.org/10.1016/j.neuron.2020.01.026

McDougle, S., & Hillman, H. (2025). Motor working memory. *Trends in Cognitive Science*. https://doi.org/10.1016/j.tics.2025.08.011

Nagel, T. (1974). What is it like to be a bat? *The Philosophical Review*, 83(4), 435–450.

Oakely, S. (2004). *Cognitive development*. London: Routledge.

Owen, A., Coleman, M., Boly, M., Davis, M., Laureys, S., & Pickard, J. (2006). Detecting awareness in the vegetative state. *Science*, 313(5792), 1402. https://doi.org/10.1126/science.1130197. PMID: 16959998.

Panksepp, J. (2014). *Affective Neuroscience: The Foundations of Human and Animal Emotions* (1st ed.). Oxford: Oxford University Press. ISBN 978-0-19-517805-0.

Plutchik, R. (2002). *Emotions and life: Perspectives from psychology, biology, and evolution*. Washington, DC: American Psychological Association.

Quiroga, R. (2012). Concept cells: The building blocks of declarative memory functions. *Nature Reviews Neuroscience*, 13, 587–597. https://doi.org/10.1038/nrn3251

Ralph, M., Jefferies, E., Patterson, K., & Rogers, T. (2016). The neural and computational bases of semantic cognition. *Nature Reviews Neuroscience*, 18. https://doi.org/10.1038/nrn.2016.150

Ridderinkhof, K., & Brass, M. (2015). How kinesthetic motor imagery works: A predictive-processing theory of visualization in sports and motor expertise. *Journal of Physiology-Paris*, 109, 53–63.

Ryle, G. (1949). *The concept of mind*. London: Hutchinson.

Sacks, O. (2012). *Hallucinations*. London: Picador.

Savage Rumbaugh, S. (2022). *Kanzi: The ape that understands and know over 3000 words*. https://www.youtube.com/watch?v=ENKinbfgrkU. Accessed on 2 April, 2025.

Savage-Rumbaugh, S., & Lewin, R. (1994). *Kanzi: The ape at the brink of the human mind*. New York: Wiley. ISBN 978-0-471-58591-6.

Sharwood Smith, M. (2004). In two minds about grammar: On the interaction of linguistic and metalinguistic knowledge in performance. *Transactions of the Philological Society, 102*(3), 255–280.

Spaak, E., & Wolff, M. J. (online, 2025). Rapid connectivity modulations unify long-term and working memory. *Trends in Cognitive Science, 2686.* https://doi.org/10.1016/j.tics.2025.02.006

Storm, K., Reiss, L., Guenter, E., Clar-Novak, M., & Muhr, S. (2023). Unconscious bias in the HRM literature: Towards a critical-reflexive approach. *Human Resource Management Review, 33,* 3. https://doi.org/10.1016/j.hrmr.2023.100969

Strimpel, Z. (n.d.). *Pets aren't people.* https://bbc.in/3cGVG83? fbclid=IwAR2wI22U18WSOuqZDPldIICmDRxgbOqVOSh_-7CM56Lczogt8XzfOJHZLKM. Accessed on 17 March, 2025.

Sutton, V. (2009). *SignWriting basics instruction.* London: SignWriting Press.

Terrace, H. S. (2019). *Why chimpanzees can't learn language and only humans can.* Columbia University Press. https://doi.org/10.7312/terr17110

Tomasello, M. (2025). How to make artificial agents more like natural agents. *Trends in Cognitive Science, 29*(9), 783–786. https://doi.org/10.1016/j.tics.2025.07.004

Tommasi, L., & Laeng, B. (2012). Psychology of spatial cognition. *WIREs (Wiley Interdisciplinary Reviews) Cognitive Science, 3,* 565–580. https://doi.org/10.1002/wcs.1198

Truscott, J. (2022). *Working memory in the modular mind.* New York: Routledge.

Truscott, J., & Sharwood Smith, M. (2019). *The internal context of bilingual processing.* Amsterdam. John Benjamins.

Xu, G., Mihaylova, T., Li, D., Tian, F., Farrehi, P. M., Parent, J. M., Mashour, G. A., Wang, M. M., & Borjigin, J. (2023). Surge of neurophysiological coupling and connectivity of gamma oscillations in the dying human brain. *Proceedings of the National Academy of Science.* https://doi.org/10.1073/pnas.2216268120

Young, M., Edlow, B., & Bodien, Y. (2024). Covert consciousness. *NeuroRehabilitation, 54*(2), 23–42. https://doi.org/10.3233/NRE-230123

# Short Glossary

*The definitions in this glossary only describes only how these terms are used in this book.*

**Activation**        The process whereby different types of *knowledge representation* and the *schemas* that connect them are brought into action ('online') on a given occasion and in response to a given task or set of tasks.

**Affective** (A)        Used for *representations* of basic emotions and *values* and of the system which stores and manages them online.

**Attention**        An enhanced state of *awareness* driven and shaped by a particular goal, which determines the object(s) towards which it is directed and its increased degree of activation.

**Auditory** (AU)        Related to the *sense* of hearing.

**Awareness**        Also called *consciousness*. An experienced state dependent on high levels of *activation*.

**Conceptual**        Related to *representations* of meaning and the *system* which stores and manages them online.

**Consciousness**        Also called *awareness*. 'Higher' consciousness is the state associated with thought processes and requires extra high levels of activation.

**Feelings**        Feelings are *sensory* experiences like taste and touch sensations and include the physical experiences of different emotions.

**Gustatory** (GU)        Related to the *sense* of taste.

**Grammar**        For any language, its (intuitively acquired) grammatical structure is created from particular associations formed between *auditory* and/or *visual* representations on the *outer ring* and *phonological representations* that are also associated with *syntactic* *representations* on the *inner ring*. The *syntactic representations* are in turn associated with particular conceptual representations also on the *inner ring*. Informal example for the verb 'kick': **/kik/** ↔verb↔strike using leg.

**Inner ring**    The location of the *conceptual, affective, spatial, phonological, syntactic* and *motor systems*.

**Pathway**    A fixed route connecting the *stores* of two *systems* through which *representations* in one *store* may become associated with *representations* in another *store* (pathways also called 'interfaces')..

**Internal Context**    The internal context is the individual's own current internal representation of the current external context.

**Motor** (MO)    Related to the *representations* of movement.

**Olfactory** (OL)    Related to the *sense* of smell.

**Online**    Related to the activity taking place millisecond by millisecond during mental processing.

**Outer ring**    Where the *visual, auditory, somatosensory, olfactory* and *gustatory systems* are located.

**Perceptual**    Related to any of the five *senses* and the five *systems* on the *outer ring* that receive information originating in the external environment.

**Phonology** (P)    One of two *systems* that together shape *grammar* and which creates and organises *representations* which have been formed via associations with given *perceptual representations* on the *outer ring* (notably auditory and *visual* ones).

**Primitives**    *Representations* that are already provided in a *store* at birth and therefore do not have to be created from experience. They are the building blocks from which all other representations in a particular store are formed and constitute human biological inheritance for each system.

Representations    Chunks (items) of knowledge created each using the special code of their *system* and managed during online processing. They include all kinds of knowledge, sometimes permanently hidden from awareness and hence not only knowledge which we come to know and use consciously.

Resting    The *dormant* (inactive) state of a *representation* or *schema* that admits of varying degrees ('resting levels'). These dictate how well they perform competitively during *online* processing.

Schema    An associative network formed from *representations* located in different *stores*. At *any given moment, schemas like representations themselves can be either in an activat(ed) or inactive state.*

Sensory    Refers to how both the brain and mind process what is registered by one or other of the sense organs (ears, ears, nose, tongue etc).

Somatosensory (SO)    Used of *representations* of various bodily sensations such as touch, temperature, pain, body position (proprioception) as well as internal physical states (interoception) and the system which stores and manages them *online*.

Spatial (SP)    Relating to the system on the *inner ring*, which stores and manages spatial *representations online*.

Store    A major component of each mental *system* which stores and manages its *representations online*. Stores are linked by a fixed network of connecting pathways, allowing associative *schemas* to be formed between their representations.

Syntax (S)    One of two (syntactic and phonological) *systems* that together shape *grammar* and which governs how words and parts of words may be combined to form sentences in any language.

**System**    One of the 11 major organisational components of the mind. Each of them specialises
in a specific type of mental task. Each system includes its own unique operating principles and a *store* where its *representations*
are created and managed *online*.

**Value**    Is a basic representation in the *affective system*, indicating either a positive or negative
state. During *online* processing, *value* associations with *schema* representations in
other *stores* can quickly change based on the
current *internal context*.

**Visual** (VI)    Related to the sense of vision (seeing) and
the system which stores and manages visual
*representations* online.

**Working Memory**    Refers to a state in which given *representations* in any *store* are currently *activated*.
There is no separate working memory store.

# Index

Pages in *italics* refer to figures.

3D space 48, 101–102

activation 15, 30, 36, 38, 41–42, 49–70, 72–75, 78, 86, 92, 95–97, 99, 102–103, 110, 112, 115–121, 123–124, 126–128, 130–132, 146, 148, 153, 159, 173; dormant 125; inactive 29, 37, 41, 50–51, 54, 68–69, 125, 175; level of activation 56
addiction 91
affective 6, 34, 46, 59, 68, 82–90, 92–97, 103, 107, 112–113, 122, 131, 162, 167, 174, 176; affective hub 85; value 46, 48, 59–61, 68, 70–71, 82–84, 86, 88, 90–91, 94–96, 103, 122, 137, 159, 161, 163–165, 176
age 63, 72–73, 93, 145, 170
AI 135
Albuquerque, N 168
amygdala 163
anaesthesia 125
anger 86, 96, 122, 137
animals 17, 34, 78, 101, 107–113, 120, 133–134, 136, 165, 169
anthropomorphism 108
appraisal 88–89, 94
articulation *154*

Ashburner, J. 171
Asimov, I. 168
association 18, 20–21, 44, 46, 48, 50, 55, 64, 70, 78, 81, 83, 85–86, 88, 90–91, 95–96, 120, 126, 128, 151–152, 158
attention 3, 49, 95–96, 111, 115, 119, 122, 124, 127, 130, 162
auditory 6, 16, 42, 78, 95–96, 100, 113, 121, 147–153, 173–174
autobiographical memories *see* memory
autonomic 10
awareness 22–23, 26, 35, 37, 49, 54, 58, 67, 92, 95, 97, 107, 110–113, 115–118, 120–127, 129–131, 133, 135, 146, 160–164, 171, 173, 175; reflective awareness 124

Baars, B. 168
Barrett, L. 8, 168
Barzykowski, K. 169
basal ganglia 104
Bechara, A. 168
bias 55–56, 65–66, 89, 172
biological 1–2, 4–5, 8–10, 16, 25, 27, 29, 37, 40, 42–43, 45, 48, 52, 67, 89, 107, 109–110, 112–113, 119,

132, 139–140, 149,
157–161, 163–165, 175;
biological hardware 2;
biological software 2, 8, 16,
52, 107, 110, 158;
biological starter 40,
42–43, 45, 48, 89, 119, 149
Black, J. 168
Bodien, Y. 172
body language 135, 137
Boly, M. 171
bonobo 77, 138, 141
Boomer, D. 168
boost 60, 70, 86, 94, 153
Borjigin, J. 172
Bortolato, T. 169
brain 1–5, 9–12, 14, 16–17,
19, 23, 25–27, 29–31,
35–36, 41, 43, 47, 49–52,
61–62, 65, 67–68, 76, 83,
88, 90–91, 93–94, 102–104,
112, 115, 121, 123, 125,
127–130, 133, 135,
157–158, 162–164,
168–170, 172, 175; imaging
30, 128–129, 133
Brass, M. 171
Brown, W. 169
Byrne, R. 170

candidate 52, 55, 58, 61,
65–66, 92
Carey, B. 168
Carruthers, P 168
cerebellum 104
Changeux, J. 169
Chomsky, N. 168
Cichy, R. 170
Clar-Novak, M. 172
Clyne, M. 168
code 40, 139, 141, 147, 175
cognitive 2, 6, 26–27, 50, 56,
66, 69, 73, 76–77, 81, 83,
92, 132, 157, 160, 169

cognitive science 26, 170–172
coherence 23, 36, 53, 71
Cohn, N. 168
Coleman, M. 171
Comatose 127
communication 75, 79,
134–137, 139, 141, 143,
146, 149, 153, 155, 162,
164, 169
competition 18, 22, 50–52,
55–56, 58–61, 67, 69, 73,
153, 159
complexity 2, 15, 39, 44, 77,
83, 110, 136, 159–160
conceptual 20–21, 39–41, 51,
53, 58, 64, 66, 68, 71,
75–88, 92, 96–99, 101–102,
104, 113, 119, 121–126,
128, 130–131, 133–137,
140, 142, 144–146, 148,
153–155, 160, 162–163,
174; conceptual hub 123;
conceptual principles 75,
137, 146; conceptual
representation 51, 64, 78,
82, 123, 148; conceptual
system 39, 58, 75–84, 88,
92, 96, 99, 101–102, 104,
113, 123, 133, 135–137,
140, 142, 144, 146, 148,
153–155, 160, 163
conscious: subconscious 17,
19, 21–23, 29, 34, 56–57,
61–62, 65–68, 75, 87–89,
97, 115, 117, 127, 142,
146–148, 158, 162,
164–165
consciousness 2, 28, 36,
54–56, 58, 65, 77, 107,
109–111, 115–117,
119–122, 125, 128,
132–133, 135, 144, 162,
164, 166, 168, 172–173;
conscious control 22, 24,

58, 66; conscious rule-learning 141; higher consciousness 133, 166
consonants 147–149
context 5, 41, 53, 55–56, 59–60, 72, 75, 80, 86, 89, 94–95, 98, 102, 104, 117, 120, 122, 128, 144, 153, 159, 172, 174, 176; external context 174; internal context 174; situational 144–145
cortex 11, 93, 102–103, 128
Crockford, C. 169

Damasio, A. R. 168
Davis, M. 171
De Bot, K. 168
De Waal, F. 114, 169
declarative *see* knowledge
Dehaene, S. 169
dissociative amnesia 68
DNA 8, 43, 89, 139, 158
Dodier, O. 169
dominant 17, 59, 96, 165, 167
dopamine 91
dream 29, 35–36, 116–117, 125–127, 131
drives 97, 161; *see also* goal
Duckworth, A. 170
dynamic 160

Eckman, P. 169
Edlow, B. 172
effortfulness 68, 124, 130
EI 158
Emo, B. 170
emotion 82–88, 90, 92–93, 95–97, 113, 122–123, 159, 167; complex emotions 96
energy 23, 52, 61–62, 67–68, 123–124, 130
enrich 101, 123
equilibrium 23

evolution 8, 43, 45–46, 93, 138, 157, 167, 169, 171
expert 19
explicit knowledge *see* metacognitive

Farrehi, P. 172
fear 6, 48, 84–87, 90, 95–96, 122–123, 141
feeling of wrongness 97, 132
Fitch, W. 169
forgetting 8, 63–64, 69–71, 74
Forkel, S. 169
Frackowiak, R. 171
framework 5, 42, 74, 102, 115, 142–143, 155, 157, 161
Frederici, A 169
Freud, S. 169
Frith, C. 171

Gadian, D. 171
genome 71
gesture 137, 139
ghost 25, 35
Girard-Buttoz, C. 169
Gladwell, M. 169
goal 160; goal-based schemas 160
goal-based schemas *see* schema
Goldacre, B. 169
Good, C. 171
GPS 103
grammar 80, 134, 140–143, 145–147, 149, 156, 171, 174–175
Graumann, M. 170
Green, A. 168
Guenter, E. 172
Guo, K 168
gustatory 16, 41, 46, 53, 121, 174

Hagoort, P. 169

hallucination 125
Handley, I. 169
Hanslmayr, S. 170
hardware *see* biological
Haupt, M. 170
Heraclitus 61
heterarchical 160
Hillman, H. 171
Hobaiter, C. 170
*homo sapiens* 43
homunculus 161
Howard, L. 170
hub 75–76, 78, 94, 123
hypnosis 70, 125

identical twins 14
Immanuel Kant 36
instinctive 24, 26, 44–49, 91,
    97–98, 119, 142, 165
intentions 65, 155, 164
interconnectivity 17, 121
involuntary 10, 105, 112, 169
Ishikawa, T. 170

Jackendoff, R. 170
James, W. 170
Javadi, A. 170
Jefferies, E. 171
Johnsrude, I. 171
Jones, S. 170
Josselyn, S. 170

Kable, J. 170
Kahneman, D. 170
Kaltenbach, C. 170
Kandel, E. 170
Kanzi 138–140, 148, 170–171
Kerszberg, M. 169
Klein, W. 170
Knight, R. 170
knowledge 1–2, 9, 13–19, 21,
    26–29, 36–42, 45, 47,
    50–51, 57–60, 64, 66, 69,
    71, 74, 76–77, 81, 84, 101,
    103, 106, 110, 112, 117,

119, 122, 125, 130–131,
    145, 147, 159, 171, 173,
    175; autobiographical 65,
    74, 131, 169; conceptually
    based knowledge 131;
    declarative knowledge 57;
    encyclopaedic knowledge
    131; explicit knowledge 131;
    meta knowledge 145;
    procedural knowledge 57–58
Kolibius, L. 170
Kurzban, R. 170

Laeng, B. 172
Langer, E. 170
language 17
Laureys, S. 171
Laver, J. 168
law of the jungle 18, 159
LeDoux, J. 170
Lewin, R. 171
Li, D. 172
limbic 90, 112
linguistic system *see* grammar
locked-in syndrome 128
Loftus, E. 170
long term memory *see* memory

Maguire, E. 171
Mashour, G. 171–172
McDougle, S. 171
*medial temporal cortex see*
    cortex
memory 15, 42, 57, 59,
    62–68, 71–74, 119, 121,
    131, 133, 171–172, 176;
    capacity 68, 159, 167;
    episodic 63, 65, 131; false
    memories 63, 70–71; global
    working memory 121;
    short term 64; suppression
    68; working memory
    63–64, 67–68
metacognitive 57, 130–131,
    147

metaphor 3, 6, 10, 47, 54–55, 60, 158
Mihaylova, T. 172
Mills D. 168
Moss-Racusin, C. 169
motor 44, 46, 49, 91, 94, 97, 99–102, 104–107, 112, 128–129, 150, 153–154, 171, 174
Muhr, S. 172
multitasking 61
Myers, J. 170

Naccache, L. 169
Nagel, T. 114, 171
neurological 10
neuron 171
neurotransmitter 91
nocebo 163–164
noticing 7, 119

Oakely, S. 171
olfactory 16, 41, 51, 83, 110, 121, 150, 174
online *see* processing
optography 30–31
Owen, A. 171

Panksepp, J. 171
Parent, J. 172
Patterson, K. 171
perceptual 13, 28, 67, 77–78, 82, 84, 99–101, 104, 109–110, 116–117, 119–123, 126, 128, 135, 148, 150, 170, 174
Perdue, C. 170
phonological 134, 146–154, 173–175
Pickard, J. 171
Pinelo Silva, J. 170
placebo 163–164
Plutchik, R. 171
prediction 161
predispositions 47–48, 162

primates 77, 84, 107, 109, 138, 144
primitives: building blocks 174; primitive schemas 44, 46
procedural *see* knowledge
processing 10–13, 15, 18–19, 22–23, 37–38, 41, 50–51, 56–59, 62, 65, 67, 69, 75–76, 79, 81, 91, 96, 102, 109–110, 120–122, 126, 129–131, 136–137, 142–144, 148, 150, 152, 158, 162, 169–172, 174–176
psychome 43

Quiroga, R. 171

Ralph, M. 171
reality 6, 14, 28, 32, 35–36, 55, 85, 88, 99–100, 104, 117–118, 126–127, 160, 163–165
Reiss, L. 172
*representation* 2
Ridderinkhof, K. 171
Rogers, T. 171
Ryle, G. 171

Sacks, O. 171
Sackur, J. 169
Savage Rumbaugh, S. 171
Savage-Rumbaugh, S. 171
Savalli, C. 168
Savalli, C., Otta, E. 168
schema 29, 39, 41–42, 46, 50, 53, 56, 60, 68–73, 75–76, 86, 91, 94–95, 98, 121, 123–124, 152–154, 161, 164, 175–176; goal-based 160–161
self 161; metaself 161; selfhood 161–162
sensory perception 28, 51, 128

Sergent, C. 169
settings 47–48
Sharwood Smith, M. 171–172
short term *see* memory
sign language 136, 141,
    150–153
skill 58, 106, 129, 147
Smith, J. 169
somatosensory 16, 34, 44, 97,
    100, 102, 121, 150, 162,
    174
Souchay, C. 169
Spaak, E. 172
spatial *see* system
speech 139, 147–149,
    151–153
Spiers, H. 170
store 9, 13, 15, 17–18,
    20–21, 29, 37–38, 40–44,
    46, 48, 53–55, 57–60, 64,
    66, 68, 75–76, 78, 81–83,
    85–87, 95–96, 101, 112,
    121, 126, 128, 136, 145,
    147–148, 152–153, 162,
    174, 176
Storm, K. 172
*strategy* 161
Strimpel, Z. 172
suppress *see* memory
Sutton, V. 172
syllable 148
syntactic 81, 134, 146–148,
    153–155, 173–175
system 6–8, 10–16, 18–19, 24,
    26, 28–29, 34, 38–40,
    42–44, 46, 48–49, 51, 61,
    76–77, 79, 81–85, 87–88,
    90–97, 99–106, 110, 112,
    123, 127, 135, 140,
    145–155, 158–160, 162,
    167, 173, 175–176; expert
    2, 6, 13, 23, 76–77, 132,

142; spatial 46, 48,
    99–104, *105*, 107, 112,
    128–129, 153–154, 170,
    172, 174–175

Teng, S. 170
Terrace, H. 172
theory of mind 161
Tian, F. 172
Tip-of-the-Tongue 60
Tomasello, M. 172
Tommasi, L. 172
transcendental 165
Tree of Knowledge 165
Truscott, J. 172

value *see* affective
visual 9–16, 19–21, 28–32,
    40, 44, 48–49, 53, 71, 78,
    81, 94–95, 99–100, 110,
    113, 121, 128, 130,
    136–137, 149–153, 168,
    173–174, 176
voluntary 104, 106, 112, 128
vowel 148

Wang, M. 172
Wilkinson, A. 168
William James 5
Wittig, R. 169
Wolff, M. 172
working memory *see* memory
working memory capacity *see*
    memory

Xu, G. 172

Young, M. 172
Yu, Y. 170

Zaccarella, E. 169
Zisch, F. 170

For Product Safety Concerns and Information please contact our EU
representative GPSR@taylorandfrancis.com
Taylor & Francis Verlag GmbH, Kaufingerstraße 24, 80331 München, Germany